AGEING AND POVERTY IN AFRICA

W0081261

For Noy

Ageing and Poverty in Africa
Ugandan livelihoods in a time of HIV/AIDS

ALUN WILLIAMS
University of Queensland, Brisbane, Australia

Routledge
Taylor & Francis Group

LONDON AND NEW YORK

First published 2003 by Ashgate Publishing

Reissued 2018 by Routledge
2 Park Square, Milton Park, Abingdon, Oxon OX14 4RN
711 Third Avenue, New York, NY 10017, USA

Routledge is an imprint of the Taylor & Francis Group, an informa business

Copyright © Alun Williams 2003

Alun Williams has asserted his right under the Copyright, Designs and Patents Act, 1988, to be identified as author of this work.

All rights reserved. No part of this book may be reprinted or reproduced or utilised in any form or by any electronic, mechanical, or other means, now known or hereafter invented, including photocopying and recording, or in any information storage or retrieval system, without permission in writing from the publishers.

Notice:
Product or corporate names may be trademarks or registered trademarks, and are used only for identification and explanation without intent to infringe.

Publisher's Note
The publisher has gone to great lengths to ensure the quality of this reprint but points out that some imperfections in the original copies may be apparent.

Disclaimer
The publisher has made every effort to trace copyright holders and welcomes correspondence from those they have been unable to contact.

A Library of Congress record exists under LC control number: 2002027781

ISBN 13: 978-1-138-72631-4 (hbk)
ISBN 13: 978-1-138-72628-4 (pbk)
ISBN 13: 978-1-315-19148-5 (ebk)

Contents

List of Figures

List of Tables

List of Text Boxes

Preface

My interest in the lives of elders in developing countries stems from my earlier work in rural northeast Thailand, where, in 1994, I was one of a team researching home care of the terminally ill (Bennett *et al.* 1994, Williams *et al.* 1996). During this research I became particularly interested in the problems faced by the aged, who were frequently responsible for the delivery of health care in the home, at a time when they were likely to be disadvantaged themselves as a result both of their decreasing abilities as they grew older and the limited support offered to them by family, community and state.

When dealing with this situation, elders were constrained by the social and economic environments within which they led their lives. For example, their health beliefs did not concur with those of the designers of health services, and, as a result, their use of these services was inappropriate and sub-optimal. Some poor families were unable to purchase health care, others found themselves impoverished through having done so. An exploration of these environments was clearly a prerequisite to a full understanding of their behaviour when faced with a crisis. Time constraints prevented the inclusion of such an exploration in the Thai research, but I resolved to make this central to any future work.

I was fortunate in being provided with a site the research reported here by the Medical Research Council (UK) Programme on AIDS in Uganda (MRC Programme). The AIDS epidemic began in Uganda in the early 1980s, and its impacts have therefore been experienced by its elders for over a decade. Since 1989, the MRC Programme has been monitoring the progress of the epidemic amongst a rural population, which in this volume I refer to as "the MRC cohort". Research in Kikole (the name is a pseudonym), a community close to, but not included in the population comprising the MRC cohort afforded me an opportunity to identify and study the multiple impacts of the epidemic, and the social and economic contexts within which they were experienced, both retrospectively and prospectively.

Acknowledgements

I wish to thank the staff and students of the Tropical Health Program, University of Queensland. In particular, Ann Larson, Judith Fitzpatrick and Lenore Manderson, Joan Bryan, Elizabeth Bennett and Will Parks, whose thoughtful contributions and encouraging comments have helped to sustain my enthusiasm throughout the past four years.

I must also thank Karen Peachey of HelpAge International, London, for securing financial support for the project, and Henry Lubwama of the Uganda Reach the Aged Association in Kampala for his help while I was in Uganda. In Uganda, the staff of the Medical Research Programme on AIDS provided me with a fieldsite, valuable advice and essential support throughout my time in the country. From Entebbe, I must thank Jimmy Whitworth, Dilys Morgan, Brian Richardson, Lucy Carpenter, and Paul Kasozi-Kazenga, and from Masaka, Robert Pool and Marjolein Gysels. I must also thank the Whitworths, the Pools, and, especially, Brian, Sue, Laura and Amy Richardson for their hospitality – your friendship made my time in Uganda so much more enjoyable.

Among the MRC staff in Kyamulibwa particular thanks are due to Anatoli Kamali, Teddy Nakimera, Elizabeth Kabunga, and Fatuma Ssembajja. To the many others; drivers, gardeners and watchmen, who welcomed me into their community and assisted me in countless ways during my time with them, I send my thanks and good wishes. Thank you all very much.

My debt to my research assistant and good friend Tumwekwase Grace is immeasurable. His many professional skills were invaluable, while his unfailing good humour, quick wit and generosity of spirit sustained us both through many long, and sometimes difficult, days.

I must also thank my brother and sister, Julian Williams and Angela Croston, and their families, for their continuing interest and encouragement. My friends helped in so many ways. I cannot mention you all, but I particularly want to thank Martin Ashby, John Atkinson, Denis Barker, Steve Clarke, Nigel Dudley, Barb Ford, Lorna Mason, Joanne Robinson, Fernando Salazar, Russell Simmons, Maggie Simmons, Sue Stolton, and Joy Taylor. I cannot overstate the value of my wife Noy's love and support during the preparation of the manuscript.

Thanks are also due to the Cambridge University Press for permission to reproduce Figure 1.1, and to James Currey Ltd for the use of Figure 1.2.

Finally, but most importantly of all, I have to thank the participants in this study. Despite their poverty, which far exceeded my expectations, and in the knowledge that they would not benefit from their participation, every aged resident of Kikole took part in this study. In making themselves and their families available to me on many occasions, they opened their homes and their hearts to me. Their dignity and strength will be with me always.

Glossary

Arabica	Variety of coffee. Little grown in Buganda.
Baganda	The name of the tribe whose home land is Buganda. [Sing: *muganda* - a single member of the tribe].
Bataka	An individual who has freehold tenure of land.
Bayaye	*Wasters*, layabouts.
Bigenge	Leprosy. Also the name of the sickness that results from eating one's totem.
Bitteete	A variety of grass which grows in fertile soil. It is cut and placed, for comfort, on the mud floors of houses.
Bogoya	A variety of sweet banana which is eaten fresh and uncooked.
Buganda	The home lands of the Baganda.
Bulungi bwansi	Compulsory community work [lit: "For the good of the country"].
Buntubulamu	A broad concept implying the possession of courtesy, compassion, good breeding, and culture. The instillation of *obuntubulamu* was the objective of traditional, *kiganda* education.
Busukko	Sores on the legs. [Lit: the herb which is placed on the road and cause sores to erupt on the legs of those who step over it.]
Busuulu	A land tax. Calculated on the size of one's land holding. Abolished by Idi Amin in 1975.
Ddodo	Wild spinach. Eaten when other food is in short supply.
Dduuka	A small shop.
Gomborola	Sub-county. An administrative division of local government.
Gomesi	The traditional dress worn by Baganda women. It remains widely used today.
Kabaka	The chief of the Baganda tribe.
Kakayira	A small, bitter, egg-plant which grows wild in Kikole and is widely eaten. Said to be good for the gall-bladder.
Kanzu	The traditional garment worn by Baganda men. Now worn only on ceremonial occasions.
Karanami	A style of hand-woven mat.
Kibanja	Customary tenure of land. Used also to describe the land over which one has this tenure [pl: *bibanja*].
Kiganda	In the Baganda style. Used to refer to indigenous beliefs, practices, or objects.

List of Abbreviations

AIDS	Acquired Immune Deficiency Syndrome
GDP	Gross Domestic Product
HIV	Human Immunodeficiency Virus
LC	Local Committee
MRC	Medical Research Council
NGO	Non-Government Organisation
PRA	Participatory Rural Appraisal
PTA	Parent-Teacher Association
Sh.	Shilling, the Ugandan unit of currency. In 1996/7 one United States dollar could buy approximately Sh1000
TASO	The AIDS Support Organisation
TFR	Total Fertility Rate
UN	United Nations
UNAIDS	United Nations AIDS Programme
UNDP	United Nations Development Programme
UPE	Universal Primary Education
URAA	Uganda Reach the Aged Association
WHO	World Health Organisation

Chapter 1

Ageing and Poverty in Developing Countries

The lives of many older people in developing countries are dramatically, and, most frequently, negatively, influenced by the social and economic changes that accompany the development process. Migration to cities by young rural adults in Africa undermines the traditional extended family system, as the new migrants establish nuclear families in the cities while the older generation are left to fend for themselves in rural areas (Khasiani 1987, Gaisie 1989, Hunt 1989, Udvardy and Cattell 1992). Similarly, the status of the elderly is threatened by the growth of individualism and a desire for independence and autonomy, as education provides younger people with some of the skills needed to strive to obtain the Western lifestyle displayed by the mass media (Nahemow 1979, Kendig 1987, Ingstaad et al. 1992).

The HIV/AIDS epidemic in sub-Saharan Africa magnifies the problems that elders face as a result of these changes. Those dying from HIV/AIDS are predominantly younger, economically active adults and children (Nunn et al. 1994). Fewer adults are available to grow food, to earn the money that is increasingly required as the cash economy displaces the rural peasant economy, to manage community organisations and to care for the sick, for the frail elderly, and for children (Tout 1989, Prebble 1990, Ankrah 1991, Barnett and Blaikie 1992).

Women, especially older women, are particularly vulnerable to the combined effects of socio-economic change and the HIV/AIDS epidemic. Caregiving and providing for the sick and for children is traditionally a woman's task, undertaken within the supportive context of an extended family (Tout 1993, du Guerny 1997). Following the changes in family composition described above, elderly women are increasingly becoming caregivers to their sick and dying adult children (Rwezaura 1989) and surrogate parents to their orphaned or abandoned grandchildren (Hunter 1990, Dunn 1992).

This situation must be considered in the context of the demographic transition. Reductions in fertility and child mortality and an increase in life expectancy have been observed in many countries throughout the world (Keyfitz and Flieger 1990, Kinsella 1997), and as a result the number of older people alive is increasing dramatically. By 2020, for example, the numbers of people aged 60 years and over will be more than six times the number that were alive in 1950, and the number of those aged 80 years and over is predicted to have increased up to tenfold by the same date (United Nations Population Division 1994). One effect of this change is to increase the requirement for the provision of services to meet the needs of an increasing proportion of aged people.

This is at a time when the proportion of younger people who are called upon to generate the resources with which to provide these services, is decreasing. The ageing of their populations is thus creating economic problems for countries experiencing this demographic transition (Okojie 1988, Adamchak 1989, Myers 1992).

In Uganda, life expectancy remains low, infant mortality and fertility remain relatively high, and therefore the proportion of elders among the population remains low – in 1991, only 3.3% were aged 65 and over (Statistics Department 1995). These figures should be seen in comparison with those of, for example, Sweden (18%), Japan (12%) and Argentina (9%) (Kinsella and Taeuber 1993). However, although the demographic transition has so far made little progress in Uganda, the absolute numbers of aged people in its population are predicted to increase dramatically in coming decades as the large cohorts of younger adults in its current population grow older. In addition, the continuing loss of young adults to HIV/AIDS will reduce the numbers of younger, productive adults available to support the aged.

Oxfam describes development as a process which aims not only to improve the circumstances of the poor, but also to empower them to manage and direct this improvement (Pratt and Boyden 1985). However, Adamchak (1989) asserts that development is taking place largely without reference to the needs of older people, while du Guerny (1997) adds that the rural aged receive even less consideration than their urban counterparts. Gorman (1995) argues that this marginalisation stems from the negative stereotypes given to older people in all modernising societies. These represent them as out of date, passive, inflexible, non-contributing members of society, whose problems need not be addressed since they have only a relatively short life-expectancy. In view of this ageism (Apt 1992), it is perhaps unsurprising that the aged are frequently seen as either irrelevant, or an impediment, to the rapid economic growth that is the goal of many developing countries and those international organisations that facilitate such growth (Treas and Logue 1986, Warnes 1994).

Poverty amongst the aged

Chambers (1995) describes eight dimensions of deprivation as experienced by poor people: poverty, social inferiority, social or geographical isolation, physical weakness, vulnerability, seasonality, powerlessness, and humiliation. The breadth of these illustrates the potential for heterogeneity of experience among the poor, and highlights the necessity, in a study such as is reported here, to look beyond mere cash poverty when exploring their lives. Although I found all of these dimensions amongst the experiences of the aged, the major concern of all was to protect themselves and those for whom they accepted responsibility from poverty and the problems that it created. This study, therefore, is essentially one of vulnerability to poverty and its effects among a group of older people. Further, amongst the poorest members of this group, who were barely able to subsist on the resources available to them, poverty manifested itself most frequently as hunger, and vulnerability as an awareness of impending hunger. Food insecurity is, therefore, a recurring theme in this book.

Poverty is a major risk of ageing in developing countries (Sen K 1994), and the World Bank has described older people and their dependents as "poor and vulnerable" (World Bank 1993a p21). Before its extent among this group can be examined, it is first necessary to clarify what is meant by "poverty". This has been given many definitions, and here I will discuss three:

Income poverty

In cash economies, where subsistence is achieved through the purchase of commodities which enable basic needs to be met, a level of income is calculated below which poverty is said to exist. Rowntree developed a poverty line which he defined as "the minimum sum on which physical efficiency could be maintained. [This is] a standard of bare subsistence rather than living. In calculating it the utmost economy [is] practised" (Rowntree 1941 pp102-103). Whilst more recent poverty line definitions have been less stringent, they continue to relate poverty exclusively to income.

Basic needs, or subsistence poverty

This definition varies according to the needs which are seen as "basic" to human life. It accepts income as one of a number of essential needs, and its development can be traced from the work of Rowntree (1941), through Drewnowski and Scott's (1966) addition of educational, recreational, security and income needs (cited in Townsend 1970), to the 1997 United Nations Development Programme (UNDP) Human Development Report, which redefines basic needs as "material requirements for minimally acceptable fulfilment of human needs" (UNDP 1997 p16). These needs include, apart from income, access to health services and education, employment and participation in the social and political life of the community.

Just as income poverty is assessed through the application of an objective, standardised poverty line, so basic needs poverty is assessed through the use of objective indicators which describe these needs. However, defining both basic needs . and income poverty is problematic when this is done by people who are not members of the same cultural group as those to whom it is to be applied. As Townsend noted: "It has always been evident that what most people would call poverty in one society would be comparative affluence in another" (Townsend 1962 p224).

Capability poverty

This makes reference to an individual's ability to lead a life not simply lacking in impoverishment, but which is valued by and valuable to the person leading that life, and by his or her community. Thus, issues such as an individual's ethnic background, lineage, fertility, age, education, and physical ability may all be relevant to capability poverty. It can be defined, therefore, only by the members of that community (UNDP 1997), and thus sidesteps the pitfalls of the definitions discussed earlier. It does not attempt, in the first instance, to relate the values of the study population to those of another population whose circumstances may be different, or those of researchers, whose knowledge of the study population may be incomplete.

Participatory methods of estimating poverty

Comparing the extent to which these conceptualisations of poverty incorporate emic values, it can be seen that the criteria of both income and basic needs poverty are unlikely to have been identified by a member of the group of people that is the subject of the study. Only in the third, capability poverty, are the criteria which describe relative wealth and poverty set by the subjects themselves. As such, it is a qualitatively different measure from both income and basic needs poverty.

To assist in deciding which method of poverty assessment is most appropriate for the current study, it is useful to review an oft-quoted research report by Jodha (1988). Jodha believes that by choosing a limited number of variables with which to estimate poverty, and then allowing these variables to have only a limited number of values, usually during the application of a quantitative survey, researchers fail to identify or examine all the factors which influence poverty in their study population. In his study of changes in the incidence of rural poverty in Rajastan between 1966 and 1984, Jodha found that the village discourse on economic status, which emphasised qualitative issues such as the availability of choice when making life decisions, and social status and patronage, was very different from that which had previously been employed by other researchers in the same community, who focused on quantitative factors such as income, expenditure and employment. When the emic indicators were applied to the study population poverty was found to be decreasing, while the etic indicators showed it to be rising.

Having raised the issue of inconsistencies between the results of research using etic and emic studies, Jodha calculated the differences between the results of previous studies which used etic criteria to obtain values for a variety of variables, and the values given to the same variables in emic studies, and found them to be between 11% and 144%. For example, household cash income estimated through etic surveys was exceeded by that obtained through emic in-depth interviewing and participant observation by 11%, while the costs of credit from institutions were reported to be 144% greater after emic than after etic studies. In each case he ascribes the inaccuracy to the failure of the etic study to take account of one or more issue-specific factors, which subsequent emic studies identified as relevant (Jodha 1988).

Jodha's conclusion, in part, has been responsible for the acceptance of the importance of community participation in social research. Recently, for example, Amartya Sen (1997) has advocated the involvement of communities as an empowering action:

> The evaluation of quality of life and of diverse capabilities call for public discussion as a part of a democratic "social choice" procedure. (p25)

Further, the UNDP (1997) stresses that any selection of criteria contains an inescapable element of judgement, and it is therefore desirable that public discussion should play a part in the selection process:

> It is very important that the standards to be used are not determined on a top-down basis, but are open to – if possible, emerge from – a participatory, democratic process. (p16)

Poverty and well-being

To avoid confusion between poverty defined as a level of income or resource deprivation and poverty that results from the multiple dimensions of deprivation discussed above, the term "well-being" is now widely used to describe the latter. Well-being is most simply defined as the experience of a good quality of life, and "ill-being" the experience of a poor quality of life (Chambers 1997). It should be noted, however, that although well-being is influenced by material wealth, it also contains social and psychospiritual factors, identified by the population to which it is applied:

> Unlike wealth, well-being is open to the whole range of human experience, social, mental and spiritual as well as material. It has many elements. Each person can define it for herself or himself. Perhaps most people would agree to include living standards, access to basic services, security and freedom from fear, health, good relations with others, friendship, love, peace of mind, choice, creativity, fulfilment and fun. (Chambers 1997 pp 9-10)

Well-being can therefore increase or decrease without any accompanying change in wealth. Thus, in attempting to estimate well-being, the identification of monetary and non-monetary factors which influence quality of life should be considered equally important.

The exploration of well-being through participatory techniques has the specific objective of enabling members of the study community to identify those factors which mediate their quality of life for better or worse. These factors can vary widely between communities, as the following examples show. Scoones (1995) reports a wealth ranking exercise in rural Zimbabwe, in which the study community identified non-monetary aspects of wealth. Most relevant to the current study, in the context of the reported declining status of the aged, is the value they placed on the status of an individual within his or her community. Seeley *et al.* (1996) found that householders of the MRC cohort said their lives benefited from factors such as having a high level of education, being part of a large social circle, and having diverse sources of income, and suffered if they experienced, for example, poor health, or if they lived alone, or consumed a lot of alcohol. Participants in Jodha's (1988) Rajastani study felt that well-being was associated with, amongst other factors, being able to wear shoes, not having to migrate to find work, and being able to live separately from one's animals. Jodha found that the 36 households whose *per capita* real income had declined by 5% or more between 1963-6 and 1982-4 were, on average, in better circumstances at the end of this period than they had been at the start, when assessed according to 37 of 38 emic indicators. Thus, despite a decrease in income, members of these households described themselves as better off.

Jodha's (1988) work reveals another valuable attribute of emic studies: the ability to explore the experiences of sub-groups within a population and obtain an understanding of the meanings they attach to well-being. Since I sought to learn about aged people's values and perceptions, which may differ significantly from those of their children or grandchildren, I took great care to allow the aged to express themselves, and to ensure that due value was attached to their opinions during subsequent data analysis.

A number of participatory research techniques have developed in response to the recognition of the need for researchers to learn from their study participants (HelpAge International 1993, Chambers 1994, 1997). These research techniques are growing in popularity and are increasingly used in Africa (see, for example, Bishop and Scoones 1994, Guijit *et al.* 1994, HelpAge International 1995a, 1995b, Norton *et al.* 1995, Thomas and Atkinson 1995). Among the proponents of these techniques, the goal of development is taken to be the maximisation of well-being as defined by the poor themselves rather than of growth or of income as defined by economists. This principle has guided the methodology of this study.

Research Methodology

The fieldsite

The research reported here was undertaken in Kikole, a village a few kilometres from Kyamulibwa, in Masaka District of Uganda. Masaka is in that part of the country known as Buganda, the traditional tribal lands of the Baganda [sing: *muganda*]. The structure of their language, Luganda, defines them as a Bantu people, and the location of their lands, between Lakes Victoria, Edward and Albert, places them among the "Inter-Lacustrine Bantu" (Seligman 1966). Figures 1.1 and 1.2 show Buganda and Masaka district in relation to the rest of Uganda.

The Medical Research Council (UK) Programme on AIDS in Uganda has a fieldstation at Kyamulibwa, about forty kilometres north of Masaka town. I was fortunate in being able to employ Mr Tumwekwase Grace, one of the MRC programme's experienced research assistants, to act as my guide, interpreter and cultural broker for the duration of my fieldwork. He is a native Luganda speaker, and lives in the region of the study. Confidentiality problems did not arise since he was not a resident of the village where this study took place. He accompanied and assisted me during almost all my discussions with the people of Kikole.

The MRC Programme's study had involved a cohort of some 10,000 individuals for seven years before my arrival. Since it was likely that this group would be affected by this experience I chose to conduct my study in a village which was not part of the cohort. I identified six villages as possible fieldsites, had meetings with the village chiefs and conducted group discussions, open to anyone who wished to attend, during which I explained the study and asked for questions and comments.

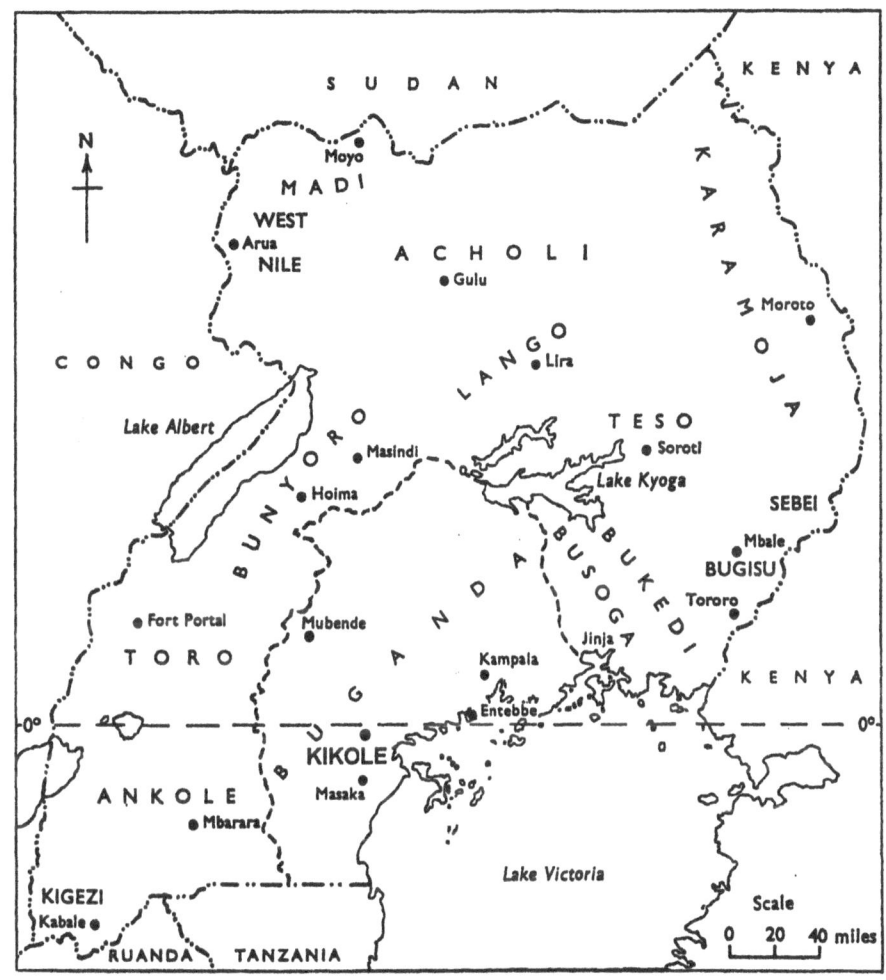

Figure 1.1   Buganda in relation to the rest of Uganda (Richards *et al.* 1973)

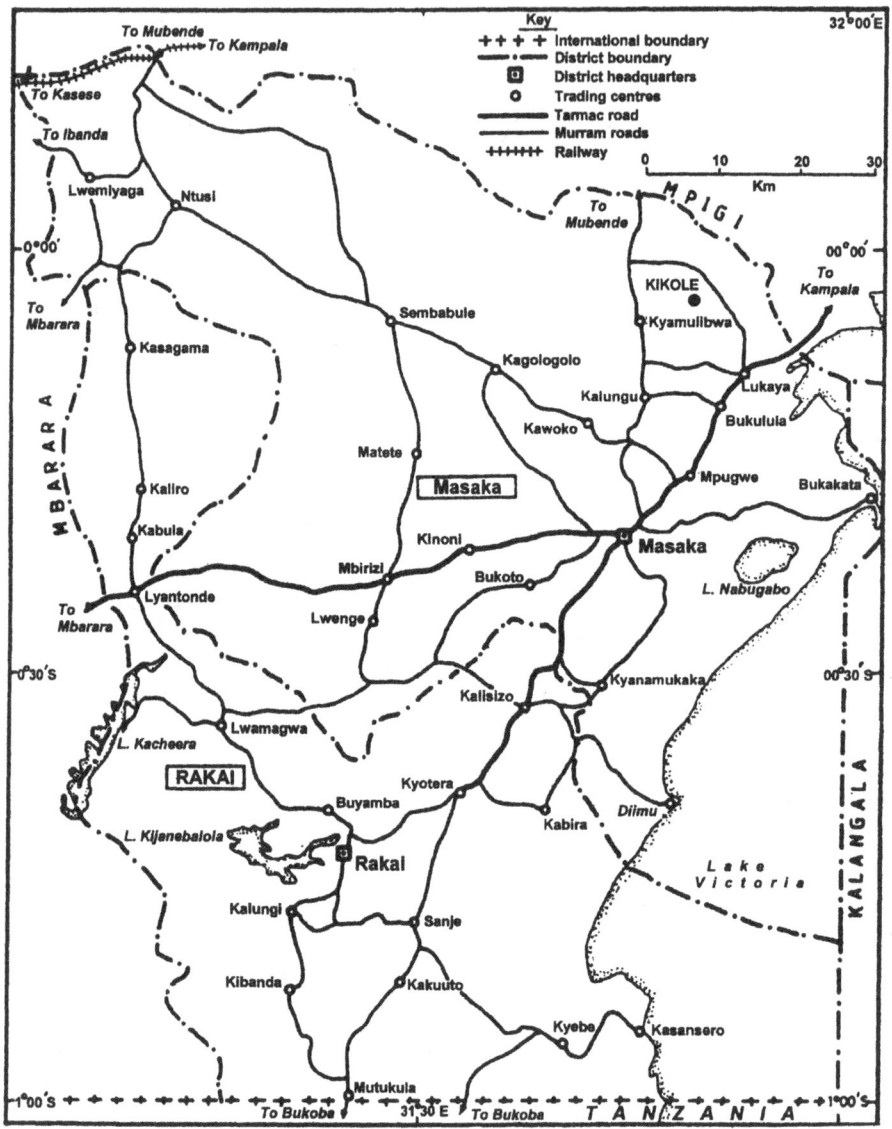

Figure 1.2   Masaka and Rakai districts of Uganda (Bond and Vincent 1981)

In two villages the people who attended the discussion were not representative of their populations as a whole, but were members of the families of the village leaders. The residents of one village were very reluctant to talk with me at all. Kikole (the name of the village and the names of its residents are pseudonyms) afforded me the warmest reception: a large number of people, both old and young, rich and poor, were eager to join a stimulating discussion which I felt was more frank than elsewhere. I therefore selected Kikole as the most suitable site for this study. The village was the most distant from the MRC cohort, and therefore the least likely to be influenced by the activities of that program. It was also the furthest from a road, and likely to be less influenced by the ribbon development taking place along the road which bordered several of the other villages I was considering as fieldsites. I therefore met the Kikole village chief a second time, explained my desire to work in the village for at least a year, and obtained his agreement to do so. At the next level of local government, I also visited the chairman of the parish council, and obtained his formal consent to my working in Kikole.

Kikole has approximately 500 residents, and is situated away from main roads. Several tracks, passable by motor vehicles in the dry season, traverse the village, but such vehicles are rarely seen as none are owned by the villagers, and the low level of business activity in Kikole presents few reasons for visits from elsewhere. Most goods are loaded on to bicycles, and pushed to their destinations by the young men who make this their living. By walking the 3-4km to the main road, which was passable in all but the wettest weather, villagers could take public transport to Kiguma to the East, or Masaka to the South, both well-serviced towns situated on the trans-Africa Highway, with rapid transport to Kampala, the capital of Uganda.

The village is situated on the top and one side of a single hill. Wrigley (1970) describes a typical Ugandan hill as having poor soil on its peak, upper and lower slopes, and good soil on its middle slopes, and this was the case in Kikole. The top and upper slope of the hill are stony and largely unsettled and uncultivated, while the lower slope is damp and also uncultivated, although where cleared it provides grazing for the village cows. Most homes are on the middle slopes of the hill, and nearly all the land around them has been cleared for cultivation. Dense plantations of coffee and bananas trees surround most houses and their compounds, rendering them invisible from even a short distance. Some houses are built on the lower slope, which, although not of great agricultural value, is seen as desirable land since it fronts the largest and most navigable track through the village, and therefore has potential as the location of a shop or other business. Their occupants cultivate strips of land stretching up to the middle slopes of the hill rather than the poor land around them. In Figures 1.3 and 1.4, which show typical village thoroughfares, the absence of visible homes is immediately apparent.

Figure 1.3   The main route through the village

The village has no electricity or telephone service, and water is drawn from springs, two protected and one unprotected, located at the bottom of the slope on which the village stands. A small number of shops and bars are located at Kilaru, the village's trading centre. Houses are mostly built on land owned by their occupants, fronted by a compound of bare earth. Here most of life's activities take place: food is prepared and eaten, laundry washed, crops dried in the sun, animals tethered, and weddings and burials conducted. Here, too, most of my discussions with participants took place. Figure 1.5 shows a relatively poor household compound.

Figure 1.4   A village footpath. The uncultivated land to the left of the path is in this condition because its residents have died and their children have left the village.

In Buganda, the traditional dome shaped house was thatched with grass from the ground to the tip of its roof, but by the 1930s rectangular houses with corrugated iron roofs had become popular (Mair 1934). In Kikole the traditional form of construction is now seen only in two kitchen buildings belonging to study participants. All the dwelling houses in the village are now rectangular, with vertical walls either of brick or wattle and daub, and pitched roofs, either thatched with grass or covered with corrugated iron sheets. Many are in very poor condition. All houses are required, by local statute, to have a pit latrine, but many of these, too, are poorly constructed and maintained.

Only wealthier households had items of furniture such as tables, chairs and cupboards in which to store possessions. Most households had little furniture, and some appeared to have none. A number of heads of households had a chair for their own use, while other household members sat on mats on the floor. It was important to be able to offer guests a chair or bench, and all but two of the households had something that served this purpose, although some households kept a clean mat specifically for this purpose, rather than a chair, and spread it on the ground for a guest. When I visited the remaining two poorest households banana leaves were picked and laid on the ground for me to sit on.

Figure 1.5   **A typical household compound.** The house is on the right, and the grass roofed building is a kitchen. Note the small amounts of a variety of crops drying in the compound.

Everyday wear was a shirt and trousers for men, and a *gomesi*, the traditional Baganda dress, for women. Whilst men's clothes could be purchased second hand and quite cheaply, a *gomesi* was an expensive item. Non-Baganda women usually wore a skirt and T-shirt, purchased second hand from local markets. Work clothes were often in very poor condition, having been repaired many times, but most people had a set of "good" clothes to wear on formal occasions such as weddings and funerals. Some men wore sandals made locally from used car tyres, but women very rarely wore shoes of any kind.

Identifying the village elders, and obtaining their consent to take part in the study

Having chosen Kikole as the study community, it was necessary to locate its aged residents, whom I hoped would agree to become the major participants in the study. In the company of one of the village leaders[1], Tumwekwase Grace and I visited each

---

[1]   Each village has a Local Committee (LC). This has nine elected members, each of whom has a specific responsibility: chairman, vice-chairman, secretary, secretary for finance, secretary for women, secretary for youth, secretary for security, secretary for rehabilitation, and secretary for information.

household and were introduced to the individual who identified him- or herself as its head. The village leader then left us, and Grace explained the project in Luganda, stressing that participation was voluntary, that withdrawal was possible at any time, and that all our discussions would be treated as confidential. As we had done with the village leaders, we stressed that we did not have funds to distribute among the village population, and that there was therefore no prospect of payment for their participation. Grace then invited questions, which I answered to the best of my ability. We then requested verbal consent from the head of each household to take part in the village census, and permission to return at a later date to talk with them again. Consent to take part was obtained from all those identifying themselves as the head of a household in Kikole.

In developed countries, where the entry into old age is often marked by retirement from employment and the receipt of support from the state, the ages of 60 or 65 years have most frequently been used as chronological markers of the beginning of "old age" (Sen K 1994). The usefulness of such a definition in a country where life expectancy at birth is much less than these figures, where functional ability may frequently decline before these ages are reached, and where there is no expectation of state support or of a period of retirement, must be questioned (Tout 1989). Apt (1988) asserts that in African cultures, age assessment is made with reference to stages of life, most commonly childhood, adolescence, adulthood, and old age. These stages were also identified by members of the study population in Uganda, and individuals were able to ascribe specific ages at which one made the transition from one stage of life to the next. However, there was no consensus on the precise ages at which such transitions occurred. I will explore these emic conceptions of old age later in this work, but will use the attainment of a self-reported age of 60 years as an etic definition of old age for the purposes of distinguishing the study participants among the study population.

The census revealed the identity and location of all those who gave their ages as 60 and over. Among these 30 individuals there were many with ages ending in a zero or a five, and many fewer whose ages ended in other numbers. This suggests that the ages given were approximate, and that many people did not know their exact age. Each of these 30 was invited to become a participant in the study. We visited them a second time, and explained our plans to visit them at home regularly over the coming year. Once again, confidentiality and the freedom to withdraw at any time were stressed, and questions were invited and responses given before their consent to participate was requested, and once again, all agreed to participate. These 30 aged residents of Kikole then became the primary participants in this study. Figure 1.6 is a map of Kikole, and shows the location of all the homes in the village. Elders' households are numbered to facilitate cross referencing whilst reading this text.

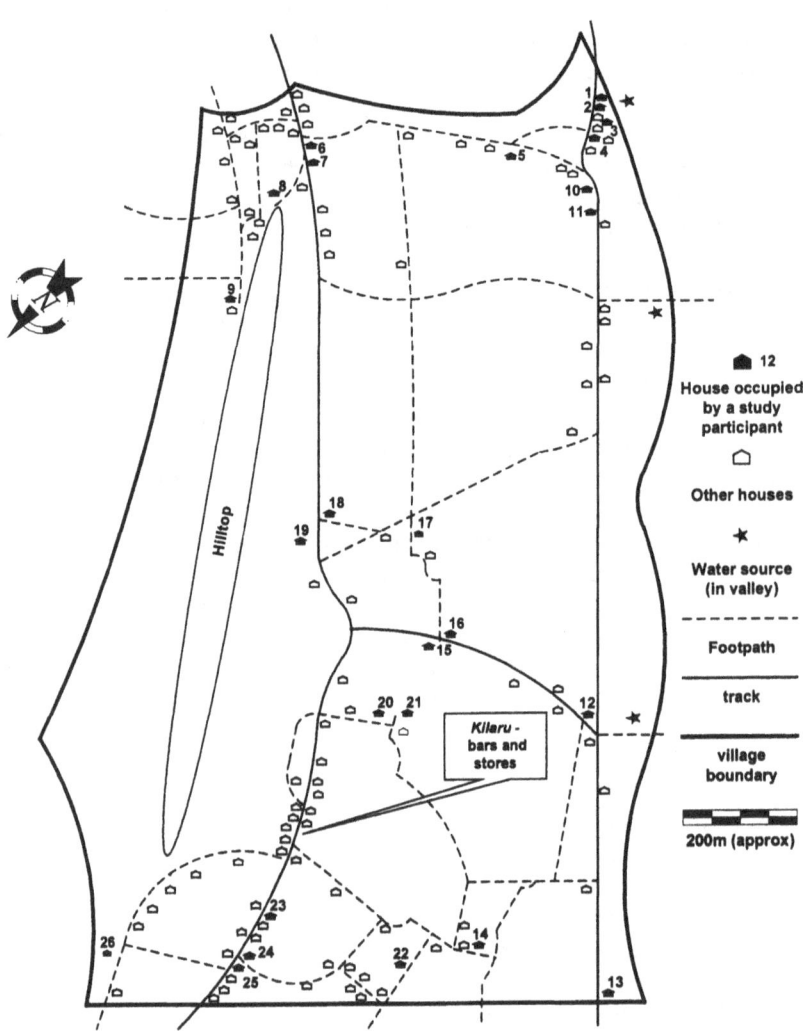

Figure 1.6   Kikole village

Collecting information

The study employed the ethnographic research approach of participant observation (Jorgensen 1989), and included in-depth interviews (Spradly 1979), the construction of life histories (du Boulay and Williams 1984), and focus group discussions (Dawson *et al.* 1993), together with techniques drawn from the rubric of participatory research, such as card sorting, wealth ranking, land use mapping, the exploration of local history, and the development of calendars describing seasonal variations in land use and diet (HelpAge International 1993, Chambers 1994).

Impromptu visits were made to each of the aged participants, at their homes, at least once a month throughout the study period. Appointments were not made, to preclude the possibility that preparations would be made for our visit. Those participants who were able spent the early hours of the day cultivating their land, and so most visits to their homes were made during the late morning and afternoon. Attempts were made, however, to visit at other times, to obtain a broader understanding of their daily routines. Discussions concerning income and expenditure, farming activity, health status, food supply and diet, and social contact took place at every visit. Time was spent following up issues raised during previous visits and in more general conversation. The development of genealogies for each household was an ongoing task, much enjoyed by the participants, and produced very useful data. At each meeting we discussed the history, location and circumstances of a section of the family, and over the year of the study I was able to acquire a comprehensive understanding of each participant's extended family support network, and of the diversity of their scale and composition. Many participants were interested by the study, and by my presence in their community, and I was often questioned closely about my life and culture. During some of these visits, throughout the year of data collection, participants were asked to perform one of a number of card-sorting and ranking exercises (Grandin 1986). Additional, single, semi-structured interviews were held with local officials whose positions gave them influence over the lives of the participants.

Further focus group discussions were held. In Kikole I conducted separate groups for men and women, both old and young. Two more group discussions took place in Kiguma, the nearest town to Kikole, and first destination of many young people migrating to urban areas. Much time was spent walking the paths of the village, moving between the participants' homes. These journeys provided many opportunities to talk with younger members of the community, passing tradesmen and visitors, and to observe the agricultural, commercial and social life of the village. The frequency with which each research method was employed is shown in Table 1.1.

Constraints

I did not speak Luganda when I arrived in Uganda, and although I rapidly learned the basics of the language, I was never able to reach a level of fluency which would have allowed me confidently to have conducted meetings and interviews in Luganda. All meetings and interviews were conducted in Luganda, most often with Tumwekwase Grace acting as my translator, although on occasions he conducted these events alone. During the few group discussions held with older women, he was replaced by two female employees of the MRC Programme, also fluent in Luganda and English. My need to communicate through an interpreter constrained the free flow of conversation during interviews and discussions, and must have limited both the quality and quantity of data obtained on each occasion. The communication problem was, to some extent, reduced by the Tumwekwase Grace's high level of interpreting skills, and eased as my language skills developed, while repeated interviews with most participants allowed unclear issues to be resolved and missing data obtained.

As a very conspicuous outsider, I was naturally treated with both interest and suspicion when I first visited Kikole. Some data collected during early discussions was contradicted by that obtained later, after I had established good relationships with the participants. They acknowledged that they had, initially, been concerned that I would pass information about their ownership of land and animals to the tax inspectors, and so had not provided honest responses during the village census. However, the long period during which I collected data allowed time for this problem to be identified. Also, as I was an outsider, my understanding of the local culture was necessarily limited, and as a result there was, most significantly at the beginning of the study, a potential for misinterpretation of data. Once again, the length of my fieldwork provided opportunities for these mistakes to be identified and corrected. In spite of the fact that both Tumwekwase Grace and I are male, I did not notice any gender related obstacles to data collection, and only during group discussions with aged women did I employ female research assistants. I did experience, however, difficulty in talking with the three married female participants, since, unless I found them at home alone, they always allowed their husbands, who were also participants, to speak on their behalf. Although I did speak to each of them, alone, on one occasion, I do not feel that I obtained as a clear an understanding of their lives as I did of those of other participants, and consequently their opinions may not be as well represented here as those of others.

Those whose mobility was restricted were unable to attend group discussions, and so did not have the same opportunities to contribute to them as more able-bodied participants. I attempted to counter this relative exclusion by spending more time with them at their homes. Among this group, those whose homes were not adjacent to one of the main paths through the village were unlikely to be able to benefit from the exchange of greetings and information with their neighbours and with passers-by that was available to those whose homes were visible from a path. Consequently, their local knowledge may have been incomplete or out of date. Some frail participants declined, on occasions, to answer questions that required a knowledge of events outside their household, citing their poor knowledge as reason for doing so. Others always answered these questions, raising the possibility that the quality of their replies was compromised by their limited, or possibly outdated, local knowledge.

Table 1.1   Research techniques employed during data collection

Research technique	Frequency of use	Location	Participants
Open discussion	6	Six potential study villages	Interested residents of six potential study villages
Focus group discussion	3	Kikole	Male elders
	2	Kikole	Female elders
	3	Kikole	Elders' sons and grandsons
	2	Kikole	Elders' daughters and granddaughters
	1	Kiguma	Young men, rural-to-urban migrants
	1	Kiguma	Young women, rural-to-urban migrants
In-depth interview	151	Kikole	13 male elders
	148	Kikole	17 female elders
Structured interview	4		Village chief, Tax inspector Priest, Traditional healer
Informal discussion	Many	Kikole and surrounding villages	Opportunistic conversations, varying in length, with villagers and others
Wealth ranking	7	Kikole	Male elders
	7	Kikole	Female elders
Age ranking	10	Kikole	Male elders
	5	Kikole	Female elders
Development of genealogies	13	Kikole	Male elders
	17	Kikole	Female elders
Land use mapping	8	Kikole	Male elders
	7	Kikole	Female elders
Participant observation	Ongoing		

Time constraints prevented younger households being studied to the same extent as the participants' households. Therefore, apart from drawing some demographic comparisons in the following chapter, I am unable to compare the circumstances of these two groups of households.

Few participants were mentally frail, and none were extremely so. I was able to have coherent conversations with everyone, but a few were, at times, unable or unwilling to sustain our conversation long enough to provide the information I was hoping to obtain. My nursing experience[2] leads me to the conclusion that this was due to mental frailty, but the possibility that they simply did not want to talk with me must be considered. Since conversation with these participants was not easy, observation played a larger part in my assessment of their circumstances than for those of other participants.

Analysing the information

I analysed qualitative data manually in Uganda, and after my return to Australia I used NUD*IST software for this task. EPI-INFO and SPSS software was employed to analyse quantitative data.

The concept of the "livelihood" and its use as an analytical tool

The lives of the old people in Kikole varied widely. One man had several wives and over twenty children, a herd of cows and over fifty acres of land, while a very old and infirm widow had a single, but disabled and ageing daughter, no animals, and a small, overgrown plot of land. With such a range of circumstances among the elders there were many contributing factors to identify and examine. Some of these, such as the climate, market conditions, the road system and the health services, were relevant to the lives of all village people, while others, such as physical disability or the loss of one's children to AIDS, were especially significant in some individual's lives. To facilitate the exploration of all these factors it is necessary to employ as wide and inclusive an analytical framework as possible.

A "livelihood" is defined by Chambers (1988) as "adequate stocks and flows of food and cash to meet basic needs", while Chambers and Conway (1992) put forward the simpler definition of a livelihood as "a means of securing a living". To date, much of the study of livelihoods has been concerned with populations experiencing food insecurity (see for example, Corbett 1988, Longhurst 1988, Frankenburger and Goldstein 1990, Frankenburger 1993, and Davies 1996). The analysis used owes much to the work of Amartya Sen. Sen states:

---

[2] I am a Registered Nurse, with extensive experience of providing health care to isolated and impoverished Aboriginal Australian communities.

Starvation is the characteristic of some people not *having* enough to eat. It is not the characteristic of there not *being* enough to eat....Whether and how starvation relates to food supply is a matter for factual investigation. (Sen A 1981 p1)

As Sen continues, if one wishes to understand the factors leading to individual starvation it is necessary to understand the factors which influence that individual's access to and ownership of food. Implicit in this statement is another acknowledgment, of the value of emic studies in the exploration of food security. Sen describes ownership as one of a number of "entitlements" of an individual, and lists other types as "trade-based", "production-based", "own-labour" and "inheritance and transfer" entitlements, which together and among others comprise the individual's "entitlement system". Sen argues that to understand poverty in general and starvation in particular it is necessary to understand the appropriate entitlement system, which, in this study, is the entitlement system of the individual aged residents of Kikole.

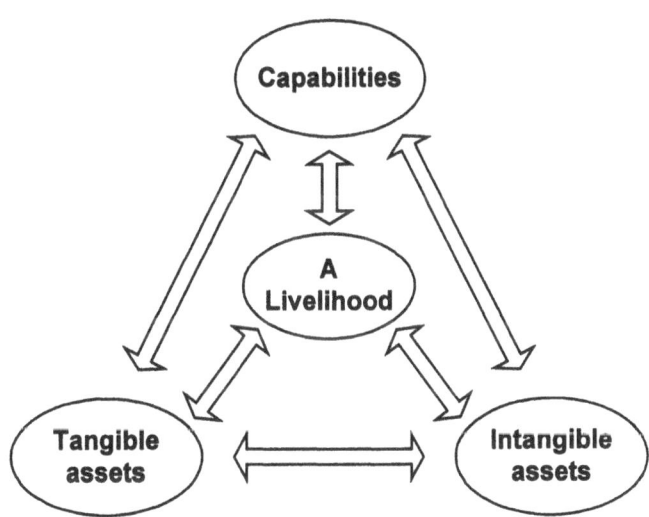

Figure 1.7   Components and flows of a livelihood (after Chambers and Conway 1992)

Food security studies tend to focus on production entitlements as early indicators of forthcoming food insecurity, and exchange entitlements (market conditions) as later indicators. Other entitlements, particularly those which are intra-household, intra-family, or in some way household specific, receive less attention. Davies (1996) attributes this imbalance to the perception that the latter data, being largely qualitative in nature, are less significant than the "harder", quantitative data concerning production and exchange, and to a consequent reluctance to accept the additional expenditures of financial and human resources associated with collecting this data. However, this group of entitlements assumes greater importance in the lives of the aged who, through their declining physical ability, are likely to experience reductions in production, and to become more reliant on other entitlements, particularly those which originate in reciprocal obligations between household and family members. It is therefore, in a study of older people's lives, not appropriate to concentrate on the more easily examined and analysed entitlements to the exclusion of others.

Chambers and Conway (1992) present a conceptual framework which divides a livelihood's resources into capabilities and assets. This is reproduced as Figure 1.7. Capabilities refer to an individual's ability to function effectively, and are dependent on physical and mental abilities, skills and aptitudes. Assets may be tangible (land, animals, stores, savings, etc) and intangible (claims rooted in the local financial, political or moral economies). It can be seen that Sen's "entitlements" and Chambers' and Conway's "assets" are broadly comparable.

Following a description of Kikole's elder population, their families and their households, I present an analysis of their livelihoods. In acknowledging Jodha's (1988) warning against a research process which initially examines only what can be easily measured, and later assumes that what was not examined was of no importance, it is my intention, in the following chapters, to give as much attention to entitlements arising from relationships with family and others, as to those arising from production and exchange.

Chapter 2

The Elders of Kikole

Within the anthropological literature on Africa two groupings predominate: those of the household and those of lineage (Guyer 1981). Both are relevant to the lives of the people of Kikole, and both will be addressed here. Bohannon (1963 p78) defines a household as "a local or spatial group, marked by propinquity" and a family as "a kinship group, marked by kinship relationships". He adds that there is no need for a family to live as a single household, nor for a household to consist entirely of members of a single family. Ageing adults will increasingly need support from others, who are not necessarily found exclusively within either household or family. This analysis does not, therefore, aim to separate and discuss the significance of lineage and household individually, but through the exploration of the relationship between them, to produce information relevant to the livelihoods of the study participants.

The household, through its propinquity, is more easily identified and enumerated than is the family, and for this reason it is usually chosen as the unit of study by census takers (Wilk and Netting 1984). But the meaning of residential propinquity is problematic, and many questions have been raised in the literature: is propinquity the same as co-residence? Do dwellings that share the same yard belong to the same household? Do single people living alone constitute a household? What about a boarding school, or a barracks? A hospital.....? Are family members who are mobile and not often in residence, but have no other home, part of the household? (Yanagisako 1979). According to Bender (1967), what distinguishes the members of a household from other co-residential groups of individuals is their shared involvement in domestic activities. Bender did not give a precise meaning of "domestic activities", but the subject has been explored further by Yanagisako (1979), who concludes that "domestic" functions are those pertaining to food production and consumption, and to social reproduction, including child-bearing and child-rearing.

Two major issues must be born in mind when using the household as the social unit in census taking. Firstly, unless the definition of "household" is developed with reference to current local norms, acknowledging both traditional behaviour and any social and economic changes that are taking place, the households identified are unlikely to resemble those that actually exist (Russell 1993). The implication here is that there is, in fact, a household to define, but this may not be the case. Preliminary research may, therefore, be needed to ensure that the definition chosen is appropriate to the community under investigation. Secondly, assuming the use of a suitable definition of "household", data gathered at the household level, whilst facilitating inter-household analysis, does not, by definition, allow exploration of intra-household relationships and resource allocation (Giesler 1993, Appleton 1996). This is an important limitation when the focus of the study is a potentially disadvantaged group such as the aged. However, inter-household comparison may show that certain types of individuals are found more or less frequently in households with specific characteristics (Drèze and Srinivasan 1995), and this method can therefore prove enlightening in the current study. The Uganda *1991 Population and Housing Census* (Statistics Department 1995 3:5) uses "a group of persons who normally live and eat together" as its definition of "household" and to facilitate comparison with national statistics, the same definition was adopted in this study.

Finally, the normative nature of the concept of household limits its usefulness in social research (Piwoz and Viteri 1985). Wolf (1966) noted the tendency of researchers to gloss over the subtleties of household composition by imposing their own pre-conceived patterns of kinship and social organisation on their data. Had the only data available to the current project been that collected during the community census then I may have fallen into the same trap. However, repeated visits to the participants' homes enabled me accurately to determine the size and composition of their families and households.

The community census

During the census, we obtained the name, age, sex, and relationship to the head of household, of all members of each household. Since much rural economic activity involves travel to urban areas for trading purposes, individuals who were not currently resident but who were present in the household for at least half the year were recorded as household members. Data were also collected concerning the construction and condition of the home, and the ownership of land, cattle, and various household items which may have been indicative of relative wealth.

The census identified 126 houses, of which 15 were unoccupied. In most cases the working definition identified households that consisted entirely of all the residents of a single house, and 101 of the 111 occupied houses were defined as households in this way. The remaining ten houses were components of four extended family groups: one group of four houses (a mother and three adult, unmarried sons who each lived in their own house near their mother, who continued to cook for them), two groups of two houses (a mother and nearby son, and a couple and their son), and of two houses headed by a single, polygamous, man. Each of these four groups was considered to function as a single household, and the total number of households in the village was thus calculated to be 105.

These households comprised a total of 494 residents, 260 (52.6%) male and 234 (47.4%) female. There were 261 children aged under 15 and 30 adults aged 60 and over, comprising 52.8% and 6.1% of the population respectively. The corresponding figures for the whole Ugandan population in the 1991 census were 47.3% under 15 years old, and 5.0% aged 60 and above, indicating that the village contained slightly higher proportions of both groups than did the whole Uganda population. This may be due to the economic outmigration of young adults that I will discuss later. The age and sex distribution of the study population is shown in Figure 2.1.

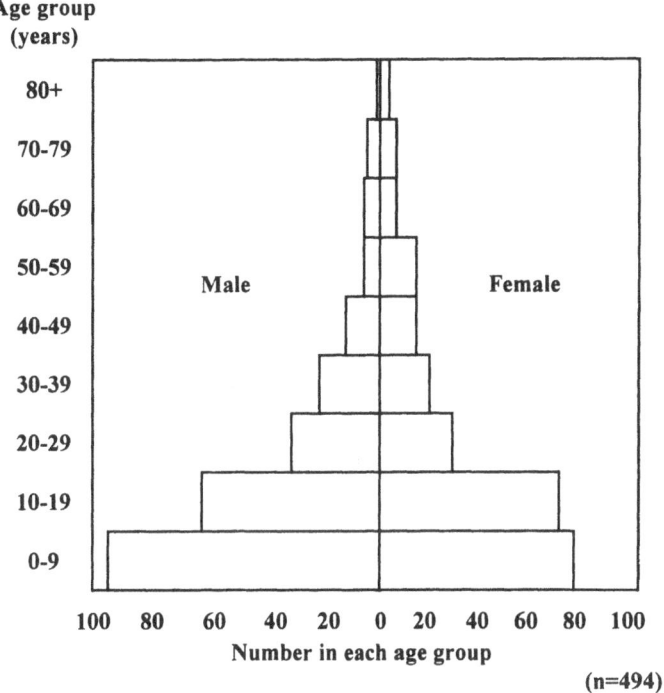

Age group (years)

Figure 2.1   Age and sex distribution of the population of Kikole

The elders, their households and families

Ethnic origin

Since the 1920s Uganda has received a steady flow of migrants from Rwanda and Burundi, the Banyarwanda and Barundi, some of whom have settled in Kikole. I shall show later that non-Baganda participants are disadvantaged in numerous ways. The ethnic origin of the study participants is shown in Table 2.1.

Marital status

In developed countries, studies have shown that old people who are married have better access to social support networks, higher longevity and health status than do those that are widowed (Gore 1990, Goldman *et al.* 1995), but comparative data on the health status of married and widowed aged men and women in Africa is not available. Table 2.2 illustrates that 12 aged men were married, compared to only three aged women. These three were married to three of the aged men, and the other nine aged men had younger wives, some of whom were still producing children. Aged women, therefore, were more likely than men to be unmarried.

Table 2.1   Ethnic origin and sex of adults aged 60+

Ethnic origin	Males	Females	Total
Baganda	8	10	18
Non-Baganda	5	7	12
Total	13	17	30

Table 2.2   Marital status of adults aged 60+

Marital status	Males	Females	Total
Married	12[a]	3	15
Widowed	1[b]	9[c]	10
Separated	0	5[d]	5
Total	13	17	30

[a] Includes three polygamous men.
[b] This man was widowed only a few weeks before the study commenced, and remained unmarried at the end of the study period.
[c] Includes one woman who had been expelled by her now deceased husband and had not remarried.
[d] Four had left their husbands and one had been expelled by her husband. All had chosen to remain unmarried.

Elders' households

Sex of the head of household   In 1992-3, the Republic of Uganda conducted a detailed household survey, in which they employed the following definition of "head of household":

> The member of the household under whose authority the activities of the household including expenditures are carried out, and who is accepted as such by all the members of the household. (Statistics Department 1992 p12, cited in Appleton 1996)

This definition of "head of household", although perhaps naïve in its simplicity, was also used during the current study. Those containing adults aged 60 years and over will be referred to as "aged households", and those not containing adults in this age range will be called "other households".

Female-headed households are over-represented among the rural poor in sub-Saharan Africa (Whitehead 1990). Appleton (1996) has analysed data produced by the *Uganda National Integrated Households Survey 1992-3* (Statistics Department 1994), and reports that household headed by widowed women are significantly poorer than other households. He asserts that this is due to lower levels of education among the heads of these households rather then to pervasive gender discrimination.

The numbers of female-headed households in Uganda has been increasing for several decades. In 1956, Richards (1966), studying a village on the outskirts of Kampala, reported that 14% of households were headed by women, and Bond and Vincent (1991), reviewing recent studies, note that 39% of all households in Masaka and Rakai districts were found to be female-headed. However, no data relating the proportion of female-headed households to the age of the head of household are available, and so it is not possible to ascertain whether or not the increasing numbers of female-headed households were evenly distributed throughout the population. In Kikole, households headed by older adults were more likely to be female-headed, as shown in Table 2.3. Table 2.4 indicates that the proportion of female-headed households amongst households headed by those aged 60+ was almost twice the proportion in households headed by younger adults.

Table 2.3   **Distribution of Kikole households by age and sex of the head of household**

Age of head of household (years)	Male- headed households	Female- headed households	All households
10-19	2	1	3
20-29	17	1	18
30-39	20	5	25
40-49	13	7	20
50-59	6	7	13
60-69	7	6	13
70-79	5	5	10
80+	1	2	3
Total	71	34	105

This proportion of female-headed households in the village is close to the national figure for rural areas of Uganda, of 72.2% male-headed and 27.8% female-headed households reported in the 1991 census (Statistics Department 1995). It is likely that the figure of 26.6% female-headed households among the under-60s in Kikole is an underestimation, since some women named their husband as head of household, even though he was working permanently in Kampala. If he continued to support his rural household financially then he would be considered the head of household according to the working definition, but as discussed later, few men did support their families in this way, in which case the wife would be considered *de facto* head of household.

Table 2.4   Number of households headed by those aged under 60 and 60+, by sex of head of household

	Male-headed households	Female-headed households	All households
Aged households	13 (50.0%)	13 (50.0%)	26 (24.8%)
Other households	58 (73.4%)	21 (26.6%)	79 (75.2%)
Total	71 (67.6%)	34 (32.4%)	105 (100.0%)

The preponderance of female headship in older households was not, as might be expected, a result of greater female life expectancy. While life expectancy at birth in Masaka district was 52.4 years for women and 47.7 for men in 1991, in 1969 a 50 year old Uganda woman (who would be approaching 80 years of age today) had an expectation of 21.03 years of life, compared to 20.72 for a man, and in 1991 an 80 year old could expect 4.85 more years of life, while a man of the same age could expect only slightly less, 4.45 years (Statistics Department 1995). The greater proportion of old women heads of households is most likely a result of their having married men older than themselves. Most reported the deaths of their husbands rather than separation from them, and their status as head of household follows from their inability or disinclination to remarry after being widowed.

Table 2.5   Religion and age of heads of households

	Roman Catholic	Protestant	Muslim	Total
Aged households	18	5	3	26
Other households	66	6	7	79
All households	84	11	10	105

Religion   Table 2.5 indicates that approximately 80% of the heads of Kikole households described themselves as Roman Catholic. The remainder consisted of almost equal numbers of Protestants and Muslims. Consideration of the elders' households alone shows that there were slightly higher proportions of Muslims and Protestants in this group, although the group, as a whole, was largely Roman Catholic.

Household size   A household's size was defined as the total number of individuals of all ages who spent at least half the year as residents of that household. This definition was intended to take account of seasonal rural-to-urban migrants and adopted or fostered children who may move between households. In Kikole, female-headed aged households were smaller than other households, as shown in shown in Table 2.6.

Table 2.6   Mean household size by age and sex of head of household

	Male-headed households	Female-headed households	All households
Aged households	7.8 (n=13*)	3.3 (n=13)	5.5 (n=26)
Other households	4.3 (n=58)	4.9 (n=21)	4.4 (n=79)
All households	4.9 (n=71)	4.3 (n=34)	4.7 (n=105)

* Number of households

The 1991 Uganda census used 65 rather than 60 years of age to separate the aged from other adults, and so figures are not directly comparable. However, Table 2.7 indicates that the census shows the same trend in household size for male- and female-headed households according to age.

In a study of rural and urban Kenya, Khasiani (1987) found that, among adults aged over 50, mean household size decreased with increasing age. In Kikole the mean size of households also decreased as the age of the head of household increased. Further, whilst households headed by women aged less than 60 were of similar size to those headed by men of that age, those headed by women aged 60+ were found to be smaller than comparable male-headed households, due, in part, to the absence of a spouse; very few old men, and the majority of old women were unmarried. Additionally women's greater age, and the longer period since they produced their last children makes it less likely that they have any of their own children living with them. Table 2.8 describes the relationship between household size, and the age and sex of the head of household.

Table 2.7   Mean household size by age and sex of head of household: all rural households in Uganda

Age of head of household (years)	Male-headed households	Female-headed households	All households
15-29	3.7	3.7	3.7
30-49	5.7	5.1	5.5
50-64	6.0	4.5	5.5
65+	5.2	3.8	4.7
Total	5.1	4.4	4.9

Source: Statistics Department (1995)

Table 2.8   Mean size of aged households in Kikole, by age and sex of head of household

Age of head of household (years)	Male-headed households	Female-headed households	All households
60-69	9.7 (7*)	5.0 (5)	7.6 (12)
70-79	6.2 (5)	1.8 (5)	4.0 (10)
80+	2.0 (1)	3.0 (3)	2.8 (4)
All households	7.8 (13)	3.3 (13)	4.7 (26)

* = number of households

Co-residence with adult children   According to the works of Gore (1990) and Goldman *et al.* (1995), the married state brings health benefits, suggesting that living with others will improve an individual's health status. While many participants were not married or widowed, most of these did not live alone, but with children or grandchildren.

Roscoe (1911) reports the existence of an avoidance relationship between parents and their married children. Breaking this taboo results in *obukko*, a debilitating and potentially fatal illness (Bennett 1963), which will be discussed further in Chapter 5. A consequence of this avoidance is that aged people are unable to co-reside with, or take food with, their married children. Reporting a 1971 survey in Buganda, Nahemow (1979) found that "the ideal living arrangement was separate from their children" (p176). Bennett and Mugalula-Mukiibi (1967) found that, in Buganda, people living alone as a result of this avoidance suffered health problems to a greater extent than members of extended families found elsewhere in Africa.

In Kikole, it is very unusual to find an aged parent living in the same household as one of their married offspring, although some adult daughters who had separated from their partners had returned to live on their father's compound. In such a case, the daughter would normally sleep in a separate hut, rather than in her parents' home, but would play a full part in the life of the household. In the following analysis daughters who are living at their parent's home in this way are treated as household members. Unmarried sons may also move from their parents' home into a separate building on their father's compound, but continue to eat with their parents and participate in the household economy. These sons are also treated as members of their parent's household in this analysis.

Table 2.9   Household composition by sex of heads of aged households

Household composition	Male-headed households	Female-headed households	Total
Couples with children	2	-	2
Couples without children	1	1[a]	2
Three generation	4	2	6
Polygamous, three generation	3	-	3
Grandparents and grandchildren	3[b]	7[c]	10
Lone adults	-	3	3
Total	13	13	26

[a] This household contained two aged sisters. It is included here since the category "couples without children" is intended to identify those who have another household member to call on for help when required, and to make a distinction between these people and those who live entirely alone, who are included in the "Lone adults" category at the bottom of the table.
[b] Includes one man whose children are adults, and who is now living with his deceased brother's wife and children.
[c] Includes a woman whose children are adults and who is now caring for her brother's young children.

The relationship between gender of head of household and household composition is explored in Table 2.9. Since men continue to produce children later in life than do women, a household headed by an aged man and which contains his non-adult children also contains a wife who is considerably younger than he, who is the mother of those children. It is likely that he will have had children with another partner earlier in life, but most if not all of these children will have matured and left home. In some instances households headed by old men with younger wives and their children also contain grandchildren produced by the children of a previous wife. In households where all children have matured and left home, the adults are likely to be living alone, or with their young grandchildren.

Co-residence with another adult While all the men living with another adult are living with their wives, all nine women who are sole adults are widows. Co-residence with another adult is shown in Table 2.10. Of the women who live with another adult, three are living with their husbands, two are widowed sisters living together, and three are living with adult daughters.

Table 2.10   **Co-residence with another adult among adults aged 60+**

Co-residing with another adult	Male	Female	Total
Yes	12	8	20
No	1	9	10
Total	13	17	30

Co-residence with younger adults Younger relatives can provide more support than an aged spouse, but since, as noted above, *kiganda* avoidance relationships prevent adults living with their children's spouses, the opportunities for co-habitation with one's children were limited to situations where these children were unmarried, or living apart from their spouse. In Kikole, three aged women were living with daughters who had separated from their husbands. Nine aged men were living with their younger wives, and of these nine, five were also living with adult, unmarried children. Table 2.11 documents co-residence with a younger adult, and shows that aged men have greater access to support from younger adults living in the same household.

Household dependency The usual measure of household dependency is the "household dependency ratio", defined as the total number of household members aged either under 15 or over 59 (considered to be "dependent" or "unproductive" individuals), divided by the number aged 15 - 59 (considered to be "independent" or "productive" household members). This measure cannot be used here, since some of the old persons' households would have zero as the denominator, giving an infinite dependency ratio, which would not allow comparative analysis. An infinite dependency ratio would also suggest that the household in question was non-viable, which was certainly not the case in Kikole. The ages chosen to separate the dependent from the independent have to be questioned in the light of the ability of households containing only the old and the young to survive with very little outside support. Further, the use of a specific age at which one becomes dependent does not reflect actuality. Ageing individuals experience a continuing change, from independence towards dependency, as age and/or frailty increases (Khasiani 1991). Similarly, children are not completely dependent until the age of 15, and some households could not survive without the inputs of much younger children. From the age of five or six years, many of these children perform essential domestic tasks which their older relatives could not perform themselves, such as collecting water and firewood and labouring on the household's land.

Table 2.11   Co-residence with an adult aged 18-59 years among adults aged 60+

Co-residing with an adult aged 18-59 years	Male	Female	All adults
Yes	9	3	12
No	4	14	18
Total	13	17	30

With these reservations in mind, I have employed an alternative measure, the "dependency rate", defined as the total number of household members aged either under 15 or over 59 years, divided by the total number of household members. In this case a household containing only the old and/or the young would have a value of one, and one with no old or young members a value of zero. Table 2.12 shows this rate according to the age and sex of the head of household:

Table 2.12   Dependency rate by age and sex of head of household

	Male-headed households	Female-headed households	All households
Aged households	0.7	0.8	0.8
Other households	0.4	0.5	0.4
All households	0.4	0.6	0.5

In view of the large numbers of children in many aged male-headed households, and the absence of other adults in most aged female-headed households, the larger rates among older and particularly older female headed households comes as no surprise.

The elders' families

Lifetime partnerships that have produced children   Marital instability among the Baganda has frequently been commented upon (Roscoe 1911, Mair 1940, Nabaitu *et al.* 1994, Adeokun and Nalwadda 1997). In Kikole this was apparent in the greater number of partnerships engaged in by Baganda men compared to non-Baganda men. The number of lifetime partnerships of the participants that have produced children are shown in Table 2.13.

Table 2.13   **Mean numbers of partnerships of adults aged 60+, by sex and ethnic origin, that have produced children**

Sex of adult aged 60+ years	Baganda	Non-Baganda	Total
Male	2.6* (n=5)	1.9 (n=8)	2.2 (n=13)
Female	1.4 (n=7)	1.4 (n=10)	1.4 (n=17)
Total	1.9 (n=12)	1.6 (n=18)	1.7 (n=30)

* Includes three polygamous men, two with two wives and one with at least three

Fertility   The total fertility rate (TFR) in Uganda remained unchanged at 7.1 live births per woman between the 1969 and 1991 national censuses, but the Ugandan Demographic and Health Survey, conducted between 1992 and 1995, has recorded a slightly lower TFR of 6.9 births. This survey found that rates were higher in rural (7.1) than in urban areas (5.0), but the low level of urbanisation in Uganda limits the impact of this change on the national figure (Statistics Department 1996a).

Male participants in Kikole, through having had more than one wife, had produced more children than had female participants. The numbers of children who had died was similar for men and women, but the men had significantly more children still alive, because they frequently had younger wives who were still producing children, or had done so until recently. Table 2.14 summarises this situation, which will be shown in Chapter 6 to be highly influential on the well-being of the study participants.

Table 2.14   Fertility and child survival among married and unmarried men
and women aged 60+

Mean no. of children produced	Male (n=12)*	Female (n=17)
Deceased	4.3	5.4
Alive	9.1	2.9
Total	13.4	8.3

* One polygamous man has been excluded from this table. His second household was in a village some distance away and it was not possible to ascertain the number of children he had produced

Clan membership amongst the elders

The Baganda claim descent from Kintu, said to have been the first man on earth (Roscoe 1911) and the first Kabaka [king] of Buganda (Southwold 1965). He met Nnambi, a woman who came "from the sky" (Kaggwa 1951, cited in Ray 1980, Ray 1991) or from a clan which was already in the country (Roscoe 1911), and produced children with her. Through undergoing a series of tasks, set by her father and her brothers, regarding courtship, marriage, and producing children, Kintu established the fundamentals of *kiganda* social structure – including those of patrilineal descent and exogamous marriage (Ray 1980, 1991), which are especially relevant to the current study.

The tribe is divided into groups of common descent, or clans. Each has a totem, which may be an animal, a fish, a bird or a plant, after which it is named and which is regarded as taboo by its members. Under no circumstance may a clan member eat the totem of that clan. The origin of the clans is not clear from the literature: Roscoe (1911) suggests that Kintu created them as a method of ensuring the survival of the totems as a food supply. Mair (1934) disagrees, but offers no alternative. Nsimbi (1964) and Kiwanuka (1971) describe a *kiganda* legend which states that before Kintu arrived in Buganda, five chiefs lived there, with their own totems, and that Kintu arrived, with his followers, whom he had already appointed as clan leaders and to whom he had allocated totems. Southwold (1965) tells a similar tale, except that he reports six original chiefs in Buganda, and adds that the arrival of Kintu is said to have taken place twenty-four generations ago. The number of clans among the Baganda has been variously estimated as 36 (Roscoe 1911), 38 (Nsimbi 1964), and 48 (Southwold 1965).

Nsimbi (1964) discussed the benefits he perceived in the clan system. These include the role the clans played in keeping order and enforcing the law, since they were able to punish transgressors, and as a promoter of collective security, since clan members were obliged to support each other when in need. Members of the same clan are forbidden to marry, since they are believed to be descended from a single ancestor, and therefore "of the same blood". This prohibition assured the physical and intellectual strength of the clan, and encouraged co-operation rather than conflict between the clans.

Table 2.15   Clan membership and ethnic origin among adults aged 60+

Clan	Baganda	Non-Baganda	Total
Antelope	1	-	1
Bird	1	-	1
Cow	2	8	10
Heart	1	-	1
Hippo	0	1	1
Lion	0	6	6
Lungfish	1	-	1
Monkey	4	-	4
Pangolin	1	-	1
Royal	1	-	1
Unknown*	-	3	3
Total	12	18	30

* These were the oldest women. Two declined to give this information, saying it was not important, and one said she had forgotten her clan

Table 2.15 shows clan membership amongst the elders in Kikole. It is of note that while eight clans were represented among the Baganda, the non-Baganda population identified themselves as members of only three. Both the Banyarwanda and the Barundi were said, by elders of these ethnic backgrounds, to have fewer clans than the Baganda, but, like the Baganda, to be patrilineal and exogamous. It is possible, however, that migrants arriving in Kikole chose to identify themselves as members of a clan already represented in the village, in an attempt to join a pre-existing support network. This action would have reduced the numbers of clans represented among the Banyarwanda and Barundi today.

Those clans that were most represented in the village were so, partly, because they were the descendants of the small number of early migrants to the area. For example, the ten members of the cow clan contained two pairs of sisters, and one group of three brothers, while the lion clan members included two brothers, and the monkey clan a father and son. The eleven Baganda belonged to eight clans and nine distinct family groups, while the fifteen non-Baganda identified themselves as members of only three clans but eleven family groups, suggesting that the groups had similar opportunities to call on family members for support, but that the non-Baganda had greater access to potential support from clan members.

The estimation of individual elders' chronological age

Since the census had revealed an apparent imprecision in the self-reporting of age, I used emic methods to ascertain the extent of this imprecision. When individuals are uncertain about their own ages, their estimations are usually arrived at after relating events in their lives to the known dates of significant events in their community's history. However, this was not a practical proposition in Kikole, as many of the participants had spent their early years elsewhere, and it proved impossible to relate the dates of the significant events of one community to those of another. Instead, a card sorting exercise was conducted to obtain the relative age of individual elders. The participants were asked to sort cards bearing the names of elders into piles, according to their perceptions of each participant's age. Since many were not literate, the names on the cards were read to them, and they were asked to indicate where they would like each card to be placed. They were given the freedom to make as many piles as they wished, and individuals chose to create between three and seven piles. After this exercise had been carried out with ten male and five female participants their individual ratings were combined, to obtain a single, emic, age ranking. In Figure 2.2 the relation between self-reported age and age ranking is displayed, and it can be seen that the trend in self-reported age approximates to the age ranking.

Comments made during the above process provided insights into emic understandings of relative age. Since most of the participants had known each other for many years, and, in some cases, all of their lives, their shared history provided points of reference from which to assess relative age:

I came here in 1957 and Amos was still a youth at that time. (Emma Kabenge)

Henry looks very old because he has grey hair, but I know he had that when he was young, so he's not as old as he looks. (Vincent Katorogo)

Age was frequently estimated with reference to physical appearance:

You become old when your face shows your age. (Rebecca Nambiru)

She looks strong so she can't be very old. (William Muwonge)

Or with reference to capability:

Katorogo must be very old because he can't fetch his own water any more. (Sulaiman Kagire)

She is still working as a midwife so she can't be old yet! (Rebecca Nambiru)

Emic age rank

Self-estimated age, years

Figure 2.2   The relationship between self-estimated age and emic age rank

An individual's appearance can also be affected by his or her circumstances:

Solomon is old because he has had a hard life, but he is not old by years. (Gilbert Kasibante)

Kasibante isn't very old, although he looks like it. That's because he's ill. (Florence Mugasha)

The emic estimation of individual elders' relative wealth

A pile-sorting exercise was also used to obtain an emic estimation of the relative wealth of the participants. The cards that had been used in the age ranking exercise were also employed here. Those elders who were willing and able were then asked to sort these cards into just four piles; those who were "rich", "very rich", "poor", and "very poor". Cards on these piles were then assigned a value of +1,+2, -1, or -2. Cards which were not allocated a pile were given a value of zero if the participant wished to place it between the "rich" and "poor" piles, or excluded if he or she was unable to place it due to a lack of knowledge of that individual's circumstances. In recognition of the temptation to under-estimate one's own wealth, all self-assessments were ignored in the analysis that followed. Ten male and five female participants performed this exercise, after which a mean of the values assigned by them to each participant was calculated. This was taken as an emic measure of that participant's relative wealth, and, when they were placed in numerical order, an emic wealth ranking (EWR) for the study participants was obtained. This ranking, in which a high number indicates relative poverty and vice versa, is displayed in Table 2.16. The table includes house numbers to permit cross-referencing with Figure 1.6. In the following chapters individual participants' statements are frequently reported *verbatim*. To enable the drawing of comparisons between the statements of relatively rich and relatively poor participants, their names will be always accompanied by their emic wealth rankings, except where to do so would breach my undertaking to maintain confidentiality.

Table 2.16   Emic wealth rankings (EWR) of study participants

Elder	House number	EWR	Elder	House number	EWR
Amos Ndawula	12	1	George Kabongo	15	16
Emma Kabenge	10	2	Gertrude Najunjju	14	17
Michael Kavuma	13	3	Vincent Katorogo	20	18
William Muwonge	19	4	Joyce Namutebi	17	19
Gilbert Kasibante	11	5	Julius Kwesige	22	20
Grace Nanteza	6	6	Restetuta Nabisere	22	21
Polly Katate	3	7	Rebecca Nambiru	4	22
Ibrahim Sabiiti	1	8	Christine Nabagareka	8	23
Noel Katusabe	14	9	Caroline Muguzi	3	24
Sulaiman Kagire	17	10	Juliet Kisaakye	24	25
Florence Mugasha	7	11	Proscovia Nanyonga	26	26
Noah Sserwadda	25	12	Lilliane Barenzi*	21	27
Solomon Nsamba	2	13	Ester Namukisa*	23	27
Ephrance Nabakka	9	14	Edith Manube	18	29
Henry Ssewanyama	16	15	Irene Mutondo	5	30

* Two women were ranked equally poor

The wealth ranking process also provided opportunities to explore emic definitions of wealth. Elders' comments indicated that there were at least two definitions currently in use. The first of these was based purely on the ownership of material assets, as shown in the following answers to the question "What makes a person rich?":

Land, a big coffee plantation, and cows. (Amos Ndawula EWR=1)

A rich man has coffee or animals like cows, or he has a job. He could have a well paid job and no other source of income since he gets a lot of money. He could be a businessman. (Rebecca Nambiru EWR=22)

It's the properties a person owns and the money. The way one dresses. If you don't dress well you can't be called rich. Owning a lot of land and cows, or having a big plantation which produces a lot of coffee. (George Kabongo EWR=16)

It depends on what he has; like cars, for example. Or on how much tax he pays. If you pay Sh8,000 and he pays Sh60,000 [Sh1000 = 1US$] where does he get that much? He could have a nice house and some rooms that he rents out. (Solomon Nsamba EWR=13)

A second group of responses to the same question reveal that, for some participants, wealth had other dimensions:

However rich you become, without children you are still poor. When you produce three children, two can be clever and one stupid. The clever ones inherit your riches and become well off themselves so that they can help you in your old age. (Henry Ssewanyama EWR=15)

If we had properties without a child we would be poor because we wouldn't have anyone to remain on our property after our death. (Vincent Katorogo EWR=18)

Many reasons. Money, friends, animals, and children. But children and friends are most important because friends can help with any problems you get. Children also make you rich in the sense that they can build you a building if you need one, buy you a bicycle, take you to hospital. I call that being rich! (Ibrahim Sabiiti EWR=8)

The properties he owns, like land and having a family; brothers and your own children and having money. But the one who has children or relatives is the richest. This is because you can have a lot of money but it may be useless. If I had money but no-one at home would I be like this? [He is disabled] I would have died long ago. Some people say that having money makes you rich, but although I need money to pay the doctor's bills, if I have no-one to take me there what use is that money? (Gilbert Kasibante EWR=5)

Seen in the context of the livelihood, these final four comments are especially revealing. Here, in addition to one's tangible, material assets, intangible assets such as relationships with children and access to health services are presented as important facets of relative wealth at a time of life when capability is declining. It is apparent that the participants' concepts of wealth, by including these non-monetary factors, are very similar to Chamber's (1997) concept of "well-being", as discussed above. The following chapters present an analysis of the livelihoods of the aged residents of Kikole. I include historical data, both cultural and economic, to allow the current circumstances of the aged to be seen within the context of their previous experience. Broadly, Chapters 3 and 4 explore their tangible assets, Chapter 5 their capabilities, and Chapter 6 their intangible assets, although the complex inter-relationships between these livelihood components, both conceptually and within individuals' experience necessitates that the boundaries between the content of these chapters remain fluid. The current AIDS epidemic has affected, and continues to affect all aspects of the participants' livelihoods. To allow an holistic analysis of the impact of the epidemic on the lives of the aged I reserve discussion of this topic until Chapter 7. Chapter 8 brings together the major findings of the earlier chapters, and presents an holistic analysis of the livelihoods of the study participants from a lifecourse perspective. Also in Chapter 8, I explore the concepts of vulnerability and livelihood insecurity, before discussing the experience of both these conditions among the aged in Kikole.

Chapter 3

Land Tenure and Usage

Land has always been the subject upon which [the Baganda] could be excited or angered. (The Kabaka 1967 p180)

It is possible, within the wide variety of cultural groups in Africa, to identify some common features in their traditional systems of and attitudes to land tenure and usage. In a subsistence economy no economic advantage was to be gained by the disposal of land, since rights in land involved rights to cultivate rather than ownership. Each individual's right to cultivate land derived from his membership of a kinship group or his political status, and this right passed to his heirs over successive generations. Those who had rights to more land than they required could allow another to cultivate it, in return for a proportion of the crop. Such a payment was seen as an acknowledgment of the relationship between landlord and tenant, and thus increased the landlord's social standing, rather than as an economic transaction. Alternatively, land could be pledged as security against a loan, with the creditor being allowed to cultivate the land until the loan was repaid. Such an arrangement may, for example, have been entered into by a man seeking to pay a bridewealth for his son's marriage (Mair 1957). Although the introduction of a cash economy has fundamentally changed land tenure practices in Buganda (Mair 1933) I found the traditional practices described above to be widely employed in 1996/7, almost a century after the Uganda Agreement of 1900, which first introduced the concept of private ownership of land to the Baganda.

The significance of land in the livelihoods of the Baganda

Land represents the final refuge for Buganda's two million people, their home and ultimate livelihood as peasant cultivators; and this they clearly understand. (West 1972)

Whilst this statement may have held true for all Baganda in 1972, population growth, land shortage, and disruption or displacement during decades of war has deprived many of the opportunity to make a living from the little, if any, land that they own. Despite the declining significance of land in the lives of many Baganda, the elders in this study continue to value their land and respect its value as a resource to be both cared for and exploited. The following idealised illustration of the traditional significance of land in Buganda is derived from my discussions with aged villagers:

A *muganda's* identity as a clan member was determined by the ownership or occupation of clan land. He was born in his father's home, and on his land, and would be nourished by the produce from that land as he or she grew to adulthood. As a young man he would then be given a part of this land to cultivate for himself and on which to rear his family. At the same time he would support his ageing parents with the produce of the land they had once cultivated for themselves. In his turn he would gradually hand over control of his land to his sons. As he aged and became less able to work, he would supported by these sons, with the produce of the same land that both he and his father had cultivated in the past. On his death he would be buried in his land, and his grave would be tended by his children, and later by their children, who would continue to occupy that land. And so the cycle would continue, with each successive generation accepting land from the previous and passing it to the next.

This is clearly an idealised representation, not least because, having settled rather than been born in the village, very few of the participants had inherited their land from their father, or from any other family or clan member. Additionally, many were not being cared for by the children to whom they had passed occupation or ownership of their land, Further, daughters and widows are omitted from this analysis. However, the representation does serve to illustrate the basis of individual attachment to land, and thus its significance to the community as a whole. It is also similar to the processes of land settlement and transfer described by earlier ethnographers in Buganda (eg Mair 1934, Richards 1966).

Today, for almost all households, land has additional significance as the generator of cash wealth, whether through the sale of food or cash crops, animals or through its rental to others. This wealth can then be used to maintain one's relationship with the state, through payment of taxes, and the purchase of health care and of education. Land also facilitates integration into the village community by producing crops or cash that may be used, by gift or exchange, to establish or maintain mutually supportive relationships with other community members, either individually through friendships and neighbourly relations, or communally, through participation in the community's ceremonies and cultural events. Farmers share the same problems of cultivation, and are vulnerable to the same vicissitudes of climate and market. Their common difficulties create an awareness of each other's problems and a mutual respect.

Thus, land has significance to older people, both as a symbolic representation of the continuation of the family from one generation to the next, that is as an affirmation of the persistence of traditional cultural values, and as the means by which one participates in the contemporary social and economic life of one's family and community. Land is the primary tangible asset of the household, the exploitation of which supplies the household with food, and allows the purchase of secondary tangible assets such as clothes, household items and dietary supplements. Further, its produce can be utilised to create and maintain intangible assets; relationships with individuals and institutions which can be called upon to provide support when it is needed. Land issues permeate much of the decision making that takes place in the village, and I therefore propose to start this analysis of the life of elders in Kikole by exploring the issues surrounding their ownership, occupation and utilisation of land.

The historical context of land tenure in Kikole

Prior to the Uganda Agreement of 1900 the Kabaka controlled land use through chiefs whom he appointed (Roscoe 1911). These chiefs administered parts of the Kingdom, controlling the use of the land over which they had authority. Land being plentiful, the right to use land was more important that its ownership, and individuals attached themselves to their chief rather than to specific parcels of land. When their chief was posted by the Kabaka to a new location his people would go with him, abandoning their land and starting afresh at their new settlement (West 1972).

The Uganda Agreement of 1900 established land as a commodity that could be bought and sold within the cash economy that was being developed and promoted by the colonial power at that time. The result was a major change in the social and economic structure of Buganda (Mair 1933). The agreement gave the Government of Uganda control of 9,000 of Buganda's estimated 19,600 square miles. The remainder, apart from a small amount given to missions and other settlements or designated as forest reserves, was divided between the Kabaka and several thousand land owners, and also amounted to 9,000 square miles. This land was allocated by area; since it was measured in square miles, or in fractions of the same, it became known as "*mailo* land" (West 1972). Although more recent legislation has changed the meaning of this term it remains current in Kikole, to denote land to which the owner has a formal, rather than customary title.

In what seems to be a lack of foresight on the part of the colonial power, aspiring *bataka* [owners of freehold land, sing: *mutaka*] were allowed to choose the *mailo* land they wanted until the limit of 9,000 square miles was reached, at which point the remainder of Buganda's land fell under the control of the government. Comparison of landholding and rainfall maps shows that *mailo* land has more rainfall than government land and is therefore more suitable for the permanent cropping system that the Baganda favour. In contrast, the government land, unwanted by those selecting their *mailo* land, was largely unproductive and relatively dry land, or infested with *tse-tse* fly or other pests which rendered it uninhabitable (West 1972). This is of continuing relevance to the villagers of Kikole, since the village's land is divided in two by the boundary between an area of *mailo* land and another of government land, suggesting that the climate may be only marginally suitable for the farming system that is practised upon it.

If, like most Baganda, one did not own *mailo* land, then one obtained the right to use land through entering into an agreement to purchase *kibanja* [plur: *bibanja*] from a *mutaka*. West (1972) defines *kibanja* as "an inheritable interest, being a small residential and cultivation tenancy, usually though not exclusively on *mailo* land." Today, elders who are holders of *kibanja* see their tenure as consisting in the right of cultivation, rather than the ownership of land, a concept which appears closely related to the pre-colonial importance of the right to use rather than to own land, discussed earlier:

> On the government land you should only dig six inches down, as below that still belongs to the government. (50 year old man)

This is *mailo* land, and belongs to Mulo. I don't have ownership of the land, just of the soil. (George Kabongo EWR=16)

The agreement with the *mutaka* involved the payment of an initial fee called a *kanzu* (Muhereza 1994), or a cockerel (Robertson 1978). A *kanzu* is the traditional long shirt or tunic worn by Baganda men, and before the arrival of the cash economy the initial payment may have been made in this form or, as more commonly in Kikole, as a cockerel:

> On government land there were two chiefs responsible for selling land....The important thing was to take a cockerel to both of them. (55 year old man)

There were also two recurring annual payments; *busuulu*, which was calculated according to the size of the *kibanja*, and *envujjo*, the size of which was related to the amount and type of produce obtained from the land:

> If your land was owned by someone you had to pay a *busuulu* of eight shillings per year, and the *mutaka* could decide how much *envujjo* you had to give him from your produce. So, if you brewed beer from your *mbidde* [small bananas from which a beer, *mwenge muganda*, is made] you would have to give him a large calabash of beer. (50 year old man)

Originally, these charges applied to food crops only, and a portion of this crop was given to the chief to support his household, but with the introduction of cash crops the *mailo* landholders, readily embracing the principles of capitalism, extended them to include a proportion of these as well (West 1972). Indeed, the *bataka* levied *busuulu* and *envujjo* at a higher rate on cash crops than on others (Muhereza 1994).

Mamdani (1976), in a Marxian analysis of the history of land tenure in Uganda, sees the owners of *mailo* land as a parasitic class whose creation by the British colonial power removed a troublesome group of indigenous leaders from the political arena. They produced nothing, but were able to make a living by renting *bibanja* to those who produced food and cash crops, a proportion of which they would then claim for themselves.

Landlords' unrestrained exploitation of their tenants caused concern within the colonial government, which introduced the *Busuulu* and *Envujjo* Law in 1928, which both placed limits on the size of these charges and gave the tenants security of tenure over their *bibanja* (West 1972), as long as they cultivated their land (Mamdani 1976) and continued to pay *busuulu* and *envujjo* as required by the law (Wrigley 1970). It was under this legal framework that the first residents of Kikole purchased their land in the late 1940s and early 1950s:

> This land was originally owned by Mugwanya, the father of Mulo. It was the Kabaka's land but was ruled by Mugwanya, who was related to the Kabaka. When Mulo sold it it was already marked out in two acre lots. We bought one of those *bibanja*, but people with money could buy land titles from Mulo. (Julius Kwesige EWR=20)

The 1975 Land Reform Decree of President Idi Amin repealed the 1928 *Busuulu* and *Envujjo* Law, and stated that all land was henceforth to be held on a 99 year lease from the state (Mamdani 1987). The objective of the law was the removal of the obstacles to the use of land for economic development purposes presented by both customary tenants and *mailo* landlords. Much land was seen as being economically unproductive, in that it was largely used for subsistence farming. If the land could be freed for larger scale farming then credit could be obtained to fund such a venture, using the land itself as collateral. Therefore, under the decree, every occupier of *kibanja* on government land became a tenant under sufferance of the state, and could be evicted, with appropriate compensation, should their land be required for other purposes, while *mailo* land was converted to state owned land and its owners became formal lessees of the state (Muhereza 1994). Mamdani (1987) states that this decree "ushered in a second flowering of landlordism in Buganda", and thus reversed the effects the 1928 *Busuulu* and *Envujjo* Law in that it served to disadvantage the holders of *kibanja*. The *bataka* realised that they too could engage in profitable trading ventures after using their leased land as security to obtain funds from banks. Land occupied by *kibanja* holders was not acceptable to the banks, and Bazarra (1994) presents evidence of forced pastoral land enclosures in Masindi district of Bunyoro (NW Uganda) during the period after the 1975 decree where the state, using the powers of that decree, and *bataka* illegally, evicted many *kibanja* holders against their will. In Buganda, where cultivation rather than pastoralism was the predominant agricultural system, Muhereza (1994) reports that to reduce the possibility of their eviction holders of *kibanja* planted permanent crops such as coffee and bananas, the presence of which on their land would have increased the compensation to which they would have been entitled upon their eviction. Consequently, there were fewer evictions of peasants in Buganda, and I heard of only one having taken place in Kikole.

Local factors may also have contributed to the reported infrequency of evictions in Kikole. Compared to many other parts of Buganda, Kikole was, at that time, in an isolated area which had an uncertain climate, and the value of *mailo* land was relatively low. In 1964-5 it was worth less than half the land to the south which, being nearer Lake Victoria had more reliable rainfall (West 1972). This would prevent the *mailo* holder obtaining, after he had compensated his *kibanja* holders for their eviction, a more than minimal net benefit from its use as security against a loan. Similarly, the government could have had little alternate use for the portion of the village land that was under its ownership. Further, the fact that the village was not settled until the late 1940s may indicate that it was not considered desirable land. Finally, the village contained several large areas of uncultivated *mailo* land that were said by the villagers to be owned by "people who live in Kampala". These people never visited the village and made no use of their land. Their land had been sold to them by the *mailo* owner, who had chosen to do this rather than evict his tenants and make use of his land, which may indicate that he felt it was of little use to him. Thus, and paradoxically, the poor quality of their land, while ensuring their continuing poverty, also provided *bibanja* holders with greater security of tenure.

The mailo land in Kikole was originally owned by a single *mutaka*, Mugwanya, who had been appointed by the *Kabaka*. In the 1960s he sold, through his son Mulo, large parts of his land. Fifty acres was purchased by one of the wealthiest men resident in Kikole, who was also a participant in this study and the only one who held *mailo* land:

> I bought land from Mulo, so my land is *mailo* land and I have the title to it.....I bought this land when I was working as a teacher. (Amos Ndawula EWR=1)

Some elders gave no indication that they felt insecure in their land tenure as a result of Idi Amin's 1975 decree:

> Some of my land is owned by a landlord and some by the government. The landlord doesn't have the right to send me away, and I can leave that land to my son as his property. But the government can tell me to leave their land if they have another plan for it. They will pay me money, including some for the value of my crops so that I can buy another piece of land. Also, on government land you had to pay ten shillings per year, and if you failed to pay you could be sent away from the land. But since the regime of Amin Dada we have stopped paying these fees to both the government and the landlords. (50 year old man)

One *mailo* owner had apparently lost interest in his land since all payments to him had stopped:

> Before Mulo died he sold my *kibanja* to Kalibala who lives in Kalungu, but he never comes to look at it these days. We used to have to pay him every year, but since the government abolished payments to *bataka* we don't pay anything. (George Kabongo EWR=16)

The settlement of Kikole

The first arrivals had the pick of the land, while later settlers had to make do with what was left:

> There was plenty of spare land here so the first settlers got the best land and they could buy more later if they wanted to, or move to better land if theirs turned out to be bad. They could sell their first land to the new arrivals...The first people here were mostly Baganda, but later many Banyarwanda came and worked for us, and after a while they decide to stay here, so we Baganda sold our land to them. Then they went back to Rwanda and brought back their brothers. (George Kabongo EWR=16)

Since the 1920s Uganda had received a steady flow of migrants from Rwanda and Burundi. Like the Baganda, the Banyarwanda and the Barundi are members of the Inter-lacustrine group of Bantu peoples whose lands extend from northern Tanzania, to the Northern shores of Lake Victoria, and whose societies traditionally were patrilineal, ruled by a paramount ruler, whose appointed chiefs controlled land use and tenure on his behalf. (Maquet 1953, Fallers 1965, Wrigley 1970). Mair (1934) noted that Baganda men were reluctant to engage in unskilled labour, preferring instead to employ people of other tribes. Work was difficult to obtain in Rwanda and Urundi, and the wages paid by the Baganda were relatively high. In the 1940s up to 100,000 of these migrants came to work as seasonal labourers in Buganda each year. They walked at least 500km from Rwanda, through Ankole, to Buganda, and along their migration routes the government provided food and shelter for them, out of concern for the people "on whom the economic well-being of the protectorate largely depends" (Uganda Protectorate 1943).

The land of Buganda was particularly attractive to these migrants since here it was legally possible for them to acquire *kibanja* on the same terms as the Baganda, and thus escape permanently from their crowded and authoritarian countries (Wrigley 1970, Fortt and Hougham 1973, Richards 1973a, Kasfir 1976), and a small proportion of the seasonal migrants settled permanently in Buganda. Between 1948 and 1959, the population of the district within which Kikole is located increased by over 100% (Richards 1973a). The first few to settle, now the oldest men in the village, said they had come to Kikole looking for a better life than they were able to lead in their home country. Some came with their families:

The Batusi were bad rulers, and we saw from the way they were treating us that we couldn't stay in that country [Rwanda]. They wanted us to dig for them so that we couldn't dig for ourselves, and if we refused they would beat us. We sold everything and brought the money with us. My parents were very old, and when we got here they were very tired, and couldn't go any further, so we stopped and I bought this land. (Noel Katusabe EWR=9)

And others arrived alone:

In Kigali [Rwanda] I was working for the priests, laying tiles, but they never paid me so I came to this area and worked at Vila Maria [a mission hospital], digging. I went to a few other places but didn't like them, and then I met a man called Kabaale who told me there was good land here, so I came to look at it with him and I liked it and decided to settle here. (Noah Sserwadda EWR=12)

This man was joined by his brother the following year:

My father had a small piece of land in Kisoro [close to the Rwandan border] and there was a severe famine. My brother was already in this village so I came here too. And I had a [clan] sister next door and her husband was a sub-parish chief. Here we had peace and something to eat. (Vincent Katorogo EWR=18)

This last is an example of the prime reason for many men's decision to settle in the village – the presence of a relative or clan member nearby. As du Toit (1990) notes, migrants to an area previously unknown to them tend to settle near one another. Similarly, Baganda settlers in the village, although not moving into an alien area, often chose to settle near a relative. This behaviour is indicative of the importance of the extended family and clan in providing help, especially at a time when, as a new settler one would have had few other people to whom one could turn to for help. For most Baganda this relative was either their father or their brother:

We chose here because it's not far from my parents....I had a brother who lived up the hill, although he lives in Bakijjulula now, and he persuaded me to come here. (Gilbert Kasibante EWR=5)

Some men, whose fathers had little land, and in an area that was already densely occupied, had no choice but to move away from them:

> My father lived and worked on the Rulayii tea estate near Masaka. He had only a very small piece of land so I left and came here in 1951. (Henry Ssewanyama EWR=15)

One of the early Baganda settlers later bought title to their *kibanja* and became a freeholder:

> At first we were only buying *bibanja* and had no right to land. When Mulo decided to sell his land title as well he sent a man called Rugobe, who came from Mugana in Mpigi district, to tell us that we could buy land title if we wanted to. I had a friend called Kalibala from Kalungu who said he wanted to buy the title to some good land, because it was possible to collect money from people who had *bibanja* on land to which you had title. I became his agent and negotiated with Rugobe for 80 acres. Kalibala brought money and we went to sign with Mulo for the land title. Then Kalibala put me in control of his land and it is still that way today. (George Kabongo EWR=16)

One participant describes the circumstances of his arrival in Kikole in Box 3.1:

Box 3.1 Arrival and settlement, c1948

I was in Kakunyu in Kalungu subcounty, close to Bakijjulula, near the swamp and the lake. A man called Polinari was living in Kikole at that time and he used to go to the lake to get fish. One day he was passing my house when his bicycle broke down. He asked to borrow a spanner but I lent him my bicycle so that his fish wouldn't go bad while he fixed his own. When he brought mine back the next day he said he could see that where I was living the land was bad, but that there was plenty of good land available in Kikole, which was still bush. He promised to take me to see it and get me some of this land, and a few days later he took me to his house which was near here. He lived with his mother, and she introduced me to the person who controlled the land in this area, a woman. She was chosen by Mulo who worked in the palace of the *Kabaka*. She had full authority to select the chiefs of this area. They were Constante in Lwannume and Sulaiman in Kikole. They took me to Kikole and to Lwannume and just here I found a small house whose owner had died. I decided to settle here, near Polinari and his mother. The land was all covered with *bitteete* [grass], and I knew that it would produce good crops. I had to pay 30 shillings to this lady for my *kibanja*.

Land tenure among widows and unmarried women

All the male elders in the village had purchased land on their arrival. Most of the women, however, had arrived in the village with their husbands, or had come to live in the village after their marriage to a man who owned land there, and settled on their husband's land. There has, though, never been any prohibition of women owning land in Buganda (West 1972), and two women had bought land and settled in the village during its first phase of settlement. The first was an educated woman who had been in employment when she purchased her land in 1957:

> I was working in Masaka and wanted a home for my children [she was a single parent]. A friend told me about this land and so I came here and surveyed it, and decided to buy it. (Emma Kabenge EWR=2)

The second bought an area of very poor land, the highest in the village and the furthest from the road and water supply, after her husband had died and her children had not allowed her to remain on her husband's land:

> I had two children, a boy and a girl. The boy was made lame by polio. When my husband died he left me some land....but my children made me leave, so I decided to come here....In this village there is a son of my sister, and he helped me find and buy this land. (Christine Nabagareka EWR=23)

Some of the earliest settlers are still alive today, and continue to live on the land on which they settled, while large areas of land are now occupied by their children. Most of these are still relatively young, but two were themselves old enough to be included in the study. These two women, daughters of one of the first men to live in the village, had grown up on their father's land and left, on their marriages, to live in the homes of their husbands. On the ending of their marriages these women returned to the village, to live on land given to them by their father:

> My father gave me this land when I separated from my husband, and I have been living here since 1968. (Florence Mugasha EWR=35)

Fathers often gave significant plots of land such as the family burial plot to a daughter in the belief that she, being married and supported by her husband, would not be in need of money, and therefore unlikely to sell it (West 1972). Florence Mugasha's sister had also been given land by her father, but was unable to make an income from it:

> I have a very small piece of land – it is my share of my father's land, and my parents' graves are on it. I can't get money from it. (Rebecca Nambiru EWR=22)

A younger man, the grandson of the father of Florence Mugasha and Rebecca Nambiru, and therefore their nephew, gave his interpretation of the context of the transfer of land between them and their father:

> Their father distributed his land with a bias against his daughters who were married at the time, and he only gave them a small piece, expecting them to get more from their marriage. But later they separated from their husbands and came back to live on their land, so they are poor today. (Young man)

Although there are no formal restrictions on the ownership of land by either men or women, in practice older women are disadvantaged when compared with older men. While all male participants owned the *kibanja* on which they lived and farmed, less than half of the women did so, as shown in Table 3.1:

Table 3.1   Ownership of land among men and women aged 60+

Ownership of land	Male	Female	Total
Does not own land	0	10	10
Owns land	13	7	20

Three of the ten women who do not own land are married and living with their husbands, on his land. One is living with her sister on the sister's deceased husband's land, and five of the other six are living on land which was owned by their deceased husband or son, ownership of which has passed to another member of his clan. The final widow is living with her daughter on land that the daughter has purchased.

The death of a woman's husband or the ending of her marriage could leave her in a precarious situation. Some, including those discussed above, had security of tenure over their homes and land, but most did not have the resources to buy their own land, and had to rely on their relatives to help them find another home:

My husband died, and his brothers took his land....they gave me a cow and I gave it to my son who sold it and bought this land, and I came and lived here too. (Edith Manube EWR=29)

I was living in Kyamulibwa with my husband. When he died my land was taken by neighbours so I decided to come here and live with my sister. (Caroline Muguzi EWR=24)

[After I left my husband] I acquired this *kibanja* from my son. His aunt, the sister of my husband bought it for him while he was in school. When he finished school he wanted to use his education and decided not to live in the village, so he let me live on his *kibanja*. Then the aunt and her son died, and now my son has died too, so I have it to myself. (Grace Nanteza EWR=6)

I came from Rwanda with my first husband. He took a second wife and she bewitched my child and it died. So I left my husband and married another man. This house and the land belonged to him, but he died and it belongs to my son Kaggwa now. I am living on his land. (Irene Mutondo EWR=30)

There is a reluctance to allow a widow formally to inherit her husband's land or property in case she then passes it, after her own death, to her own rather than her husband's clan. To avoid this event a will may instruct women who inherit land from a father or husband to return it to his clan upon their own deaths (Southwold 1956). In just one case the deceased husband had formally left his wife secure tenancy (but not ownership, which would have removed it from the control of his clan) of some of his land in his will:

> My husband bought title to five acres and died without having sold any of it. He distributed it among his sons and daughters and I have a bit with some coffee on it. When I die it will go to one of my children, and not members of my clan. (Ephrance Nabakka EWR=14)

In spite of the existence of these safeguards, and as shown in Table 3.1 above, some widows did not have formal ownership or security of tenure over any of their husband's land and were living there with the permission of his heir, who was usually their own son. Non-Baganda widows, although culturally able to inherit land from their fathers or brothers in the same way as the Baganda (Fallers 1965), were unlikely to have members of their own lineage living nearby, and so could not call on them for help. Nor could they inherit land from the families they had left behind when they came to Buganda with their husbands, since they had long ago lost touch with them. Non-Baganda women were therefore largely unable to receive land from their own clan. Notwithstanding the earlier difficulties experienced by some, during the study none of the women participants expressed any feeling of present land insecurity. Table 3.2 addresses this relative disadvantage, which is revealed, very clearly, in the patterns of women's land ownership in Kikole.

Table 3.2   Ownership of land among women aged 60+, by ethnic origin

Ethnic origin	Owns land	Does not own land	Total
Baganda	5	1	6
Non-Baganda	2	9	11
Total	7	10	17

(Chi square, $p < 0.05$)

Rural people are more likely to be poor if they own little or no land (Lipton 1988). The discussion above suggests that, while non-Baganda men are disadvantaged when compared to Baganda through owning lower quality or less area of land, non-Baganda women, particularly widows, are further disadvantaged through having restricted access to land and no access to support from their natal family. Elders' perceptions of the relative wealth of older people in Kikole indicate, as shown in Table 3.3, the significance of both gender and ethnic origin. (Once again, a low ranking indicates relative wealth and vice versa.)

Table 3.3   Mean emic wealth rankings of adults aged 60+, by gender and ethnic origin (n=30)

Sex	Mean emic wealth ranking by ethnic origin		Total
	Baganda	Non-Baganda	
Male	6.6	12.5	10.2
Female	14.3	23.0	19.4
Total	10.4	17.9	15.4

Table 3.4 reveals that unmarried, non-Baganda women elders, who were less likely to own land than others, were also perceived by the participants as less wealthy than other women:

Table 3.4   Mean emic wealth ranking of women aged 60+, by ethnic origin and marital status (n=17)

Marital status	Mean emic wealth ranking by ethnic origin		Total
	Baganda	Non-Baganda	
Married	19.0	19.0	19.0
Not married	13.5	24.0	19.5
Total	14.3	23.0	19.4

The 1995 Constitution of Uganda contains Articles intended to prohibit discrimination on the grounds of sex, and another which, in apparent contradiction, retains the institution of customary (*kibanja*) tenure without modification. It remains to be seen how the legislature will interpret these articles (Nsibambi 1996).

The passing of land to the following generation

Traditionally, although the Kabaka owned all the land of Buganda, a man had the right of use of that land via his membership of a clan, and the clan itself had control over the transfer of land after the death of one of its members (Roscoe 1911). Today, unless a landowner dies intestate (West 1972), these decisions are made by the deceased and contained in his will. In accordance with the principle of passing land on to clan members men want their land to remain within their clan after their deaths. However, they also want the land to remain within the clan in perpetuity, and so will be wary of passing it to younger clan members who might sell it or otherwise abuse their ownership. Therefore, although the use of land may be transferred from father to son while the father is alive, ownership often remains with the father until he dies, and is only transferred during the *orumbe* ceremony, during which his will is read. By retaining formal ownership the father has continuing power over his land even after he has passed its use or occupation to his children, and should they abandon the land or die, or in some way displease their father he can reclaim the land and reassign it as he wishes:

> I gave a piece of land to my grandson Ssentongo, but instead of using it he gave parts of it to his brother Kiwanuuka and another man called Kaggwa, showing me that he wasn't going to use it well. So I took control of it again, and I will give what's left of it to the grandson who lives with me now. (George Kabongo EWR=16)

> I gave land to one of my sons, but he died so I have taken it back again. (William Muwonge EWR=3)

Should a son be seen as untrustworthy, his father may decide not to leave him land at all:

> Peter will get some land, but Mukasa did me wrong by stealing my bicycle so he won't get anything. (Solomon Nsamba EWR=13)

> If my son has gone away to look for money I will look after his land for him, but if he has been sent away for stealing I will sell it and announce that he has no land. (Ibrahim Sabiiti EWR=8)

Land as a burial ground

In the past, having protected his land from future maltreatment, the owner was buried on his land after his death, and his grave tended by his descendants (Roscoe 1911). Graves may become neglected if descendants move away from their parental home, as did the older inhabitants of Kikole when they came to settle the village. Mair (1934) reported the abandonment of burial grounds in Buganda in the early 1930s, but noted that their location was not forgotten, and that from time to time a stranger would arrive and weed a small plot of ground where an ancestor was buried, even though there was no visible sign of a grave. In Kikole, the first people to die after settlement of the village had been buried in new burial grounds on the *bibanja* that had been acquired by the settlers, and these grounds had been used since that time. Figure 3.1 shows a typical family burial ground in Kikole.

Figure 3.1   A family burial ground

Box 3.2 contains the text of a will. This was written, in Luganda, on pages torn from a school exercise book. It was shown to me by Matthew, whose father had been its executor. As he was his father's heir, he had received it after his father's death. Some of the children mentioned in it are still young, and have not yet inherited their land. They are living with their mother, who is using their land. When they are of age, Matthew will oversee the transfer of their land to their ownership, and his task will be complete. The document will then be handed to Sseguya's heir, Kalema, who will be responsible for resolving any disputes that may develop over family members' tenure of its land.

Land usage

The main food crops grown in Buganda at the start of this century were *matooke* (a plantain), sweet potatoes, maize and beans (Roscoe 1911). Cotton was the major export crop in Uganda and Kikole from the 1920s until the mid-1950s. After the second world war coffee grew in popularity as an export crop, and its acreage exceeded that of cotton for the first time in 1957, after which time cotton declined rapidly while coffee growing increased correspondingly (Fortt and Hougham 1973). Cassava was not seen by Roscoe, but by 1956, 2.8% of production by acreage in Mengo and Masaka districts of Buganda was devoted to its cultivation (Fortt and Hougham 1973), and in 1994, 10.6% of land under cultivation in Uganda was producing cassava (Statistics Dept 1996b).

Box 3.2 An example of a will

4/6/86

I, Godfrey Sseguya, make this will while I am healthy. No one has forced me to do so. If I die, I will leave seven children, four boys and three girls. Of these, my heir is to be Tadeo Kalema. These are the properties he will inherit: I have given him land from the mugavu tree down to the road. Its width extends from Gonzaga's *kibanja* to that of the widow of the late Charles. I have given part of my *kibanja* below the road to two of my children and my wife. These children are Amina and Emmanuel. If their mother is no longer alive those two children should share her part. I have already given the remaining children his or her own piece of land, and they are living there now. Janet is next to the road, then Namalwa, Naluggo, and lastly Kalema. The remaining part of my *kibanja* above the road, uninherited by others, I have given to my wife. The iron sheets which I lent to Gonzaga, five (5) in number, nine feet long and 37 gauge, should be repaid to my heir. These are the properties I have. I have selected Leopoldi as a foster parent.

Those who have witnessed this will are:

1 Charles, 2 Samson, 3 Leopoldi   (three signatures)

Secretary: Peter K (signature)

I, the will maker, Godfrey Sseguya

(Thumbprint)

In 1986/7, an agricultural survey of the administrative sub-district in which Kikole was located found that 57% of its land was under cultivation, and of this area 45% was producing bananas, 26% coffee and 29% other crops (Planning Dept 1989, cited in Bigsten and Kayizza-Mugerwa 1995). In 1995 coffee accounted for 69.3% of the monetary value of Uganda's exports (Statistics Department 1996b), and in

1996/7 coffee was the only cash crop grown in quantity in Kikole. In 1996/7 the food crops most frequently grown in the village were *matooke*, sweet potatoes and cassava, with only small plantings of maize.

.All cultivation is done by hand, usually with a hoe. This implement consists of a spade-shaped blade fixed at right-angles to a long handle, and it is used to break rough ground, to dig, to plant and to weed around growing crops. A new hoe cost a minimum of Sh4000, and many households were using extremely worn implements, with blades only half their original size. However, since these shifted less soil they were easier for elders and children to use. The hoe is an essential household item and every household in the study possessed one. In my year in the village I did not once note the use of a new hoe, or the use of any wheeled or motorised vehicle in the cultivation of crops.

The food security implications of the cultivation of cash crops

Crops may be grown as food for domestic consumption or to sell to obtain cash. Pinstrup-Anderson (1983) noted that pressure to produce cash crops can threaten household food security if land normally used to grow food is diverted to the production of cash crops. The household can then become vulnerable to disadvantageous changes in market conditions, which may result in the income gained from cash crops being insufficient to purchase food equivalent to that which they replaced. Further, in many African cultures, the income from cash crops is a male resource, while it is a female responsibility to provide the household with food. Having started producing cash crops on land hitherto used to produce food, a husband will not always use the money he receives upon their sale to purchase food, while his wife has less land than previously on which to grow the food she is expected to supply to her household (Food and Agriculture Organisation 1987). Gittinger (1990) reports that the woman's task is made more difficult should her husband appropriate the household's most productive land to grow cash crops.

Boserup (1970) states that, in most African tribal communities, men participated in agriculture only by felling trees on newly settled land, after which time cultivation became a woman's responsibility, and that men's relative underemployment made them an ideal workforce for colonial powers keen to promote the growth of cash crops. Roscoe (1911) observed that it was exclusively the women who grew food in Buganda, but in the early 1930s men were seen to be cultivating food crops as well the cotton which was the major cash crop at that time (Mair 1934). Later writers have reported that food crops were grown entirely by women (Fallers 1964a, Richards 1973b, Tadria 1987). Today, this persists as a social norm:

> The father is expected to get money to provide health care, pay school fees, and solve any other problems as they occur. The wife's job is to have children, feed them, and keep the house clean. The husband only gets involved in providing food if there is a crop failure and it becomes necessary to buy food. (Young man)

Some male elders, however, said that their labour contributed to their household's food supply:

My wife digs her own plot while I am away, and I dig mine when I get back from work. We dig separately but when the harvest comes we put all the produce together since it is all for home consumption. (Solomon Nsamba EWR=13)

"A family is like a bird: each builds its own nest. An ibis builds a different nest from a heron." In my and other households women want to have their own cash. They grow their own beans separately from their husband's, and after the harvest the cash from the man's beans will be used to feed the family and the woman will keep the cash from what she has grown. And if you have given her a plot of coffee, after the harvest the money from the man's coffee is consumed by the family, and the woman keeps the cash from her plot, and buys clothes with it. (Ibrahim Sabiiti EWR=8)

In Kikole both men and women held the opinion that assurance of the household food supply should take precedence over obtaining a cash income when dividing land between food and cash crops:

If you have only a small piece of land you grow only for home consumption. (Juliet Kisaakye EWR=25)

Food crops are more important. Only after we have enough food can we think about making money. (Noah Sserwadda EWR=12)

Two factors increased the importance of this decision. Firstly, total dependence on cash crops is unwise (Longhurst 1983), since surplus food may be sold to raise cash but coffee cannot be consumed, and a crop failure could seriously affect household food security:

You can't plant coffee on all your land as you might not be able to buy food when there is a poor coffee season. (Ibrahim Sabiiti EWR=8)

Secondly, coffee and *matooke* are both perennial plants, growing in the same land for many years, and any decision to plant them could not be easily reversed:

It depends on how much land you have. If you have a lot you can plant bananas or coffee on it and plan for it to stay there for years, but if you have only a little you have to use it to grow food to eat, not coffee. (Noah Sserwadda EWR=12)

Only richer farmers could expect to make money from their cash crops, while poor farmers had to make sure they had enough to eat. The Kikole farmers mostly identified with the latter group:

Rich people plan to get money, maybe having a big plantation of *matooke* or sweet potatoes. But we just grow to eat, and only sell if there is a good crop. (Henry Ssewanyama EWR=15)

Some people make a business out of farming, using tractors. They dig to sell, but we dig to eat. (Sulaiman Kagire EWR=10)

Major crops grown in Kikole

Bananas are important to the Baganda. *Matooke*, *bogoya* and *ndisi* are eaten, fresh or cooked, while *mbidde* and *kisubi* are fermented to make *mwenge muganda*, the local beer, or distilled to a produce *waragi*, an alcoholic liquor. Multiple varieties are often mixed together in the same plantation, and they mature throughout the year, although more slowly in dry periods.

Cassava is a popular subsistence crop because it requires few skills to produce, is drought tolerant, and can give a fair yield even under adverse conditions (Prudencio and Al-Hassan 1994). Since it can be stored in the ground until required it can be a useful resource when other foods are in short supply (Longhurst 1983). Uganda farmers, in the mid 1960s, were required, by law, to produce cassava as a reserve food supply to be used in case of famine (Hougham and Sturrock 1973). The participants said they liked to grow it because it will grow in almost any soil, requires less cultivation than other crops, and will continue to produce for up to two years after planting. For this reason it is grown by elders who do not have the strength to grow other crops. It is however, not a popular food, as its bland taste and fibrous, unyielding texture are not generally liked. Those whose teeth were missing or decayed were particularly unenamoured of this crop. An epidemic of cassava mosaic virus is currently spreading south through Buganda (Thresh *et al.* 1994, Gibson *et al.* 1996), and much of the village crop is already affected, the main result being a reduction in production.

Sweet potatoes require considerable digging and later weeding, but they are relatively unaffected by pests and will continue to mature during dry weather. In addition they are easily propagated from sections of the vines that they produce, and so the grower does not need to buy seed. They are widely grown as a reliable food crop. A small area will produce a lot more food than the same area of *matooke*, and so they are specially popular with small landowners. The preparation of land for sweet potatoes is very hard work. After cropping, some vines are replanted to produce next season, and any surplus is fed to pigs and goats.

Ground nuts, beans and peas are grown for both home consumption and for sale. They are highly unreliable crops as they need specific weather conditions to produce well. Too much or too little rain will drastically reduce the crop. Beans require good growing conditions for two months and ground nuts three. They also require considerable digging and weeding. In a poor season some growers manage only to produce a crop sufficient to provide seed for the next season.

Maize was introduced to Buganda in the latter half of the nineteenth century (Langlands 1965), and has since become a major crop nationally (Statistics Department 1996b), although it was little grown in Kikole. It was sometimes planted speculatively in odd corners of cleared land, but rarely as a field crop, since it is labour intensive, requiring the clearing of land, and good rainfall over a four month period, which could not be relied upon. The first maize crop of 1997 failed due to lack of rainfall. Rarely seen crops such as tomatoes and onions require irrigation to mature, and the labour involved in obtaining water means that only very small quantities are grown, usually next to the house, where any surplus water from cooking or washing can be used to water the plants.

New varieties of crops and new farming techniques have appeared in the village, but poor farmers were unable to increase their yields through their use as they could not afford the additional expenses associated with their cultivation:

> The big beans come from America and are called *nambale*. Most people are using them these days. We are directed by the government not to allow any weeds in our coffee, and we use fertilisers - cow dung, coffee husks, grass mulching or a chemical...This is only for those who can afford it, and even if you can afford to buy it you still have to pay someone to take it to the field. Others get poorer yields..... Many people have cows but only those who can afford to buy medicines for them or to spray their land get better yields. (William Muwonge EWR=3)

Coffee is the main cash crop of Kikole. After initial success in Buganda plantations, *arabica*, the variety which fetches the best price on world markets proved vulnerable to pests and diseases (Kajubi 1985), and almost all the coffee grown today is of the lower valued *robusta* variety. A small amount of "traditional *kiganda*" coffee is also grown, less productive than *robusta*, but more tolerant of adverse weather conditions and less demanding of its soil. By growing a few trees a farmer can ensure that in bad seasons he has at least a small crop to sell, and will have some money available to meet unavoidable costs. *Arabica* coffee is not grown in Kikole.

Berries are dried under the sun. Depending on the weather, this may take up to five days, during which time the berries harden, reduce in size and become black in colour. They are spread on the compound each morning, and taken in each evening as there are heavy dew falls most nights. Bringing coffee in at night is also said to be needed to avoid its being stolen. Young children who are not at school are often responsible for drying coffee, since it is necessary to remain near the coffee and cover it in case of rain. A good berry will contain two beans, while poor quality coffee will have only one or no fully grown beans. This may be due to poor husbandry, unsuitable weather conditions while the berries were maturing, or plant disease. Poorly dried berries may become mouldy and reduce the quality of the crop, and the price obtained will suffer as a result.

Uncertainty in crop production

Kikole is situated about 10km South of the Equator, but the climatic extremes that might be anticipated in such a location are moderated by its altitude of about 1200m above sea level and its proximity to Lake Victoria. Temperatures rarely exceed 27, or fall below 15 degrees centigrade. There are two wet and two dry seasons each year. During the first there are usually violent thunderstorms and a great deal of rain, while the second is generally less extreme. Rainfall in Kikole is approximately 1000mm per year (Griffiths 1962) with the greater part falling in the first wet season of the year. Although the Directorate of Overseas Surveys (1961) rainfall map shows that the Kikole area is likely to receive at least 750mm of rain in four out of five years, residents of the village experienced its rainfall as unpredictable, both in terms of its seasonal pattern and its quantity. The same crops are grown each season, although the first season, due to its greater average rainfall, is said to produce larger crop yields.

Figure 3.2   A woman inspects her ripening coffee. Note that she has a goitre, indicating a long-term dietary deficiency of iodine.

Figure 3.3 shows a maize crop that has failed due to a lack of rain. Alterations in rainfall patterns were thought to be responsible for changes in both the timing of planting and cultivation, and the quality and quantity of the crops produced:

> We can't rely on the rain like we used to. It stops and starts, stops and starts – we should be weeding now, not planting! The beans which are being planted now will be ready in late November, but in the past we would have been weeding now, and harvesting in late October. (Ephrance Nabakka EWR=14)

> The weather has changed. In the past we used to get much more rain. Look how small the bunches of bananas are – in the past you could hardly lift them! (Grace Nanteza EWR=6)

Residents of Kikole generally perceive that yields of all crops have fallen over the years. One explanation for this change was a perception that pest damage was increasing:

Our parents used to have a lot of food, as it wasn't attacked by pests the way it is these days. They didn't use chemicals. These days the cost of chemicals can use up all the profits from the crop! (Sulaiman Kagire EWR=10)

In the past, when our land was good, our banana plantations gave us a lot of *matooke*, but that has changed and now they are very poor. Look at that patch above my compound – see how small the banana trees are? It used to be a good plantation, but look at it now! I don't know why it has declined so much – I think it may be weevils or another disease. (Amos Ndawula EWR=1)

Figure 3.3   A maize crop that has failed due to insufficient rainfall

Mamdani (1996), reporting his research in a Buganda village, blames reducing soil fertility on peasants' inability to purchase fertilisers. Some people use traditional pest and weed control methods:

Those leaves you can see in my beans – they keep pests away. (Florence Mugasha EWR=11)

When you have prepared land for sweet potatoes beans are grown first because weeds grow in the prepared land if you leave it fallow, then after the beans you plant sweet potatoes. (Henry Ssewanyama EWR=13)

And some, facing increasing poverty, are returning to these older methods:

> We used to put "*Ddudu*" [a branded insecticide] on the banana trees, but we had to buy that. Now we use ashes instead to stop weevils. Our parents used to do this, but we forgot about it for a while. (Rebecca Nambiru EWR=22)

Perennial crop production on the same land places strains on that land which, unless relieved, will reduce its fertility and thus its future productivity. Poor farmers often find it difficult to supply the necessary inputs which will prevent their livelihoods being compromised through the reduced production of their land (Frankenburger and Goldstein 1990, Bayliss-Smith 1991). Kikole farmers understood the principles of maintaining fertility by mulching land, by rotating crops or leaving it fallow, and some were able to do so:

> I use the manure from my cows to improve my coffee. (Michael Kavuma EWR=3)

> When I first had this land the soil was very red because the owner used to dig in the weeds when they were small, so that the soil never got much extra fertility. Weeds help us a lot because they contain manure. I always leave them until they are big, or even until they die and form black humus on the soil so that I get good crops. (Grace Nanteza EWR=6)

Some did not have the resources to do this:

> I have a big *shamba* [field], but poor yields. The soil is poor but I can't afford to mulch or put manure on it. (Florence Mugasha EWR=11)

Soil depletion was the most frequently expressed explanation for decreasing crop yields. Some people saw this as an inevitable consequence of long term cultivation:

> The soil has become poor and maize and cassava don't yield well....we used to get high yields of beans, but these days you get nothing.... I started tilling my soil when I was young, and now I am old, so why shouldn't the soil get old, too? (Emma Kabenge EWR=2)

> When my banana plantation gave a good crop I ate a lot of *matooke*, but these days it doesn't produce so well – I think the land is getting poor. So now we eat a lot more cassava......perhaps the top soil is being washed away by the rain, the stones are on the surface these days. (Florence Mugasha EWR=11)

The government provides an agricultural extension service, but it is has been much reduced in recent years:

> There are no people coming from the ministry any more. Since the introduction of central government they have combined the agriculture and veterinary departments in one job, but a man can't be an expert in everything! (Tax collector)

Information now reaches the village by radio:

The radio gives us a lot of knowledge, but we don't have the money to put it into practice. We wait and see if the ideas work for other people first. (Ibrahim Sabiiti EWR=8)

One elder was a member of the Uganda National Farmers Association, and obtained information on modern farming techniques at their meetings. However, membership is Sh3000 *per annum*, and he was the only older person in the village who was a member.

Much has been written on the problems that face those experiencing seasonal variation in food supply (see, for example, writers in Chambers *et al.* 1981, de Garine and Harrison 1988, Sahn 1989 and Gill 1991). In Kikole, crop production is influenced by both sunshine and rainfall, and experience has taught farmers that they cannot rely on a good crop each season. The distance to village water sources makes irrigation of land an impractical proposition, and there is nothing people can do but wait and see how the seasons unfold:

We were praying for rain, but God gave us too much and the bean pods are full of water! Now we want rain one day and sun the next. (Rebecca Nambiru EWR=22)

It's God's plan; we just plant and wait. Some years the crops can be destroyed by not enough rain and others by too much. The farmer has no control after sowing the seeds. (Solomon Nsamba EWR=13)

Even in a good season a sudden storm can destroy one's crops:

I am going to die because I am old we don't have enough food! We used to get by with beans but they were damaged by the hailstorm last week and now we won't have enough to eat! (Vincent Katorogo EWR=18)

This uncertainty in production and therefore in food supply is reflected in a variability in cash crop production. Although coffee can be relied upon to provide two crops per year, the size of the crop varies:

In a good season I can get eight to ten tins of fresh coffee, but only three in a bad one. (George Kabongo EWR=16)

Animal husbandry

Roscoe (1911) describes the widespread keeping of cows, sheep, goats and chickens in Buganda, but does not record having seen pigs or rabbits, both of which are kept today. Early this century cows, which had ceremonial significance, were kept, even by the poor. They were not killed for meat, except by the chiefs, and there were many and complex taboos governing the use of their milk. Goats were kept by every household since they could be sold in case of an urgent need, such as the payment of fines. There were no taboos attached to their rearing or consumption, unlike those surrounding sheep, as a result of which sheep husbandry was infrequent. Chickens were killed and eaten to celebrate family events, such as the creation of a blood-brotherhood, or the return of a son from war. They were also given as gifts to chiefs and to visitors.

Cattle In the early decades of this century *tse-tse* fly was a problem in Buganda and many cattle died from sleeping sickness, or from other epidemic diseases, and by the 1930s few people owned them (Mair 1934). Few cattle are to be found in Kikole today. Initially, I formed the opinion that only the wealthier farmers, who had easy access to water and large areas of grazing land, were keeping them. These men had a highly visible *kraal* [corral] where the cows were kept overnight, and after milking each morning they were driven to the fields by a herdsman, usually a family member, who would watch them all day, water them, and bring them back to the *kraal* for a second milking in the evening. Since cattle ownership is taxable, other, smaller farmers were reluctant admit to possessing them at the start of the study, but as a rapport developed between us they talked more freely, and it became apparent that several possessed cattle that were kept elsewhere. Since these farmers did not have access to grazing land in the village they kept them in *malungu;* drier, pastoral areas to the north and west of the village, where some of the local people made a living providing grazing for others' cattle. Their ownership of cattle was therefore not obvious to the observer in Kikole. During the village census conducted at the start of this study all heads of households were asked how many cattle they owned. None of those five aged heads of households who owned cattle but kept them outside the village admitted to their ownership of them at that time, and this information was only obtained during the year spent collecting data following the census. The fact that cattle ownership is taxed is almost certainly the reason why they were not prepared to acknowledge their animals to a stranger. Several aged men and one woman said they had at one time had herds of a dozen or more, but that in recent years most of them had died, and they had brought the survivors back to the village:

> I had some but they died. I got them in 1979, and moved them from Kityaba to Rwabenge. Some of them died there so I moved the rest to Kilaga. There were 11 at first, but only five came to Kilaga. They were killed by a lung infection called *kihaha*. They have continued to die, and now I have only one old cow and its calf which are still at Isaka's home in Rwabenge. Most of them died during the past year. (Noah Sserwadda EWR=12)

> I had fifteen, being cared for by herdsmen, but they all died from *kihaha*, apart from two calves. (Henry Ssewanyama EWR=15)

It was not possible for me to confirm that these deaths had occurred, and it may be that their owners, mindful of their potential tax liability, were hoping that the tax assessors would somehow hear through me that they now had fewer cattle.

Some had difficulty meeting the costs of keeping the cows elsewhere and had brought them back to Kikole:

> Sseparaga was looking after our cow for Sh1000 a month, but he put it up to Sh1500, so we moved it, and brought it back here. (Vincent Katorogo EWR=18)

Other animals Sheep, goats, pigs, chickens and rabbits were all kept, as shown in Table 3.5. Some farmers took their sheep or goats to grazing areas, whilst others kept them tethered at home. Neither sheep nor goats were kept for milk production. Pigs were left to scavenge for themselves, or given household food scraps:

> I don't tether them – they won't go far away. I give them water, and sometimes I cook cassava for them, but usually they just find their own food. (Grace Nanteza EWR=6)

A cockerel is the popular choice of food to give a guest, or to take as a gift when one is visiting. Most households kept a few chickens which scratched a living in both the compound and the house, where they were useful consumers of insect pests. Those households that cultivated beans kept their chickens in a chicken house as they could damage the growing plants and reduce their yield. Meat rabbits were kept mostly by children, who collected leaves for them and fed them in their cages.

Table 3.5   Ownership of animals and sex of head of household aged 60+

Animal	Sex of head of household aged 60+	
	Male	Female
Rabbits	1	1
Chickens	7	5
Pigs	2	3
Sheep or goats	6	4
Cattle	9	1
None	2	5

Among the elders in Kikole, women, who had relatively poorer access to land, did not own animals as frequently as men. Among men, ownership of pigs, sheep or goats was very similar to their ownership of cattle, but more women owned these smaller animals than owned cattle.

Summary

In the agrarian economy of Kikole, survival depends on the production of crops from one's land, and tenure of land is therefore central to the lives of the aged residents of Kikole. The non-Baganda were disadvantaged at the time of settlement, acquiring less land, or land of lower quality than the Baganda. Since land is considered to be the property of its owner's clan, a woman is rendered particularly vulnerable on her husband's death, since his land is likely to be left to his children rather than to his wife. Widows are, therefore, frequently dependent on their children for their access to land. In Kikole, only those few women who had established themselves as economically independent had been able to acquire ownership of land, although the comments of others, mostly living on land owned by their husbands or sons, did not indicate any feelings of insecurity arising from their lack of formal land ownership.

Decisions regarding land use require a balance to be drawn between the food and the cash needs of the household. Animals may sicken or die and cash crops may fail, or market changes render their monetary return too small to meet the household's food needs, and therefore food crops are always grown, sufficient to meet the household's needs, after which surpluses may be sold for cash. Further, crops have to be chosen that will ensure that some produce can be stored for consumption during periods between cropping seasons. In view of the village's unpredictable rainfall, the declining fertility of the soil and increasing pest and disease problems, yields cannot be relied upon, and the threat of food insecurity is always present. The following chapter will explore the participants' ability to protect themselves from this insecurity through participation in the village economy.

Chapter 4

The Rural Economy

> The peasant runs a household, not a business concern.....The perennial problem of the
> peasantry consists in balancing the demands of the external world against the peasants'
> need to provision their households. (Wolf 1966 pp2,15)

The economic development of Buganda under colonial rule

The colonial powers in Africa were concerned to develop production of commodities
for export (Wrigley 1957), and to this end they adopted one of two strategies: they
encouraged settlers to farm large plantation estates using local labour, as in Kenya
and many West African countries, or they facilitated the production of these crops
within a local, peasant economy (Bernstein *et al.* 1992). In Uganda, the creation of
profitable plantations proved problematic, since the 1900 Uganda Agreement
allowed the Baganda to assume ownership of the most productive land, and it was
necessary for prospective European plantation owners to buy or lease land from its
Baganda owners. By 1916, non-natives had acquired 50,000 acres in Buganda,
mostly through the purchase of freehold land (Wrigley 1957). Thereafter, the sale of
freehold land in Buganda to non-natives ceased by law (West 1972).

Non-native plantations established at this time suffered from a lack of labour,
since the Baganda were able to subsist relatively comfortably on their own land, and
saw no reason to replace this lifestyle with one of continuous, poorly remunerated,
hard labour. Additionally, the Baganda had little economic incentive to change their
lifestyle since the little cash they needed to pay their taxes and meet other incidental
expenses was provided by the sale of cotton, the cash crop that peasant farmers had
been encouraged to grow by missionaries and others since the turn of the century
(Wrigley 1967).

Low market prices hindered the development of coffee plantations until 1911.
Relatively small numbers flourished until after the First World War, when prices fell,
culminating in the 1920 worldwide collapse of commodity markets, and the decline
of plantation agriculture. Peasant farmers were not greatly involved in coffee
production at this time. However, without overhead costs to bear, and able to survive
on their own food production, they were less concerned with the low price, and hence
gradually increased their cultivation of the crop, replacing the non-native plantation
owners as the main producers of coffee during the 1920s and early 1930s. Peasant
farmers in Masaka district had had less success growing cotton than those in other
parts of Buganda, and, with encouragement from the *Kabaka*, started to grow large
acreages of coffee. Between 1922 and 1934 the area of native grown coffee in

Buganda increased from approximately 600 to over 21,000 acres, and Masaka district became the largest area of peasant coffee production in Uganda (Wrigley 1967).

Peasant economies

Wolf (1966) distinguishes between "cultivators" and "peasants":

> It is only when....the cultivator becomes subject to the demands and sanctions of power-holders outside his social stratum that we can appropriately speak of peasantry. (p11)

Neoclassical economic theory, with its view of an economy "without politics, power or history" (Escobar 1995 p65), is not applicable to the study of peasant economies, since peasants' decision-making may be driven by cultural rather than economic considerations (Shanin 1987, Ellis 1988). Marxian analysis, however, acknowledges the significance of the social and political context of economic activity, and has been employed in dependency theory to explore the economies of developing Latin American nations (Dos Santos 1970, Cardoso 1972, Palma 1978). Mamdani (1976) has related Marxist theory to the Ugandan situation, while others have argued that the cultural and historical differences between the situations in Africa and Latin America are evidence for its inapplicability in Uganda (Bunker 1983). Here, I will use the Marxian analytical process to investigate both the peasant economy as an institution, and the Kikole village economy at a population level. In Chapter 8, however, where economic decision-making by aged individuals is investigated, it will become apparent that neither neo-classical nor Marxian economic theory is an appropriate analytical tool when considering sub-groups within the village population.

Marxian theory predicts that social differentiation will force peasants, through their participation in and assimilation into the capitalist economy, to divide into two classes: capitalist farmers and wage labourers. Although an indigenous capitalist elite has emerged in Uganda, the peasantry continues to exist in large numbers (Mamdani 1976, Brett 1992). A number of explanations have been put forward for the persistence of the peasantry in the face of capitalist exploitation. Banaji (1976) suggests that rather than being doomed under capitalism, the peasantry is subsumed by it. Through their use of predominantly unpaid household labour, peasants are prepared to accept relatively low prices from capitalist entrepreneurs who profit through advantageous marketing of the produce. Peasants are therefore welcome participants in a capitalist economy (Harriss 1987).

Bernstein (1979) has two explanations, which he says may occur together or separately, for the persistence of peasant economies in apparent contradiction of Marxian theory. Firstly, he notes the impossibility of the emergence of wealthy farmers in a situation where all farm surplus is removed from the impoverished farming household through, for example, market inequalities or taxation. Secondly, he reports a progressive devaluation of peasant labour which occurs when market values of farm products fall, whether due to falling production costs elsewhere or falling world commodity prices. The peasant household is thus prevented from increasing its income, even though it may, if it has surplus labour, increase its production. In both cases the result is that the household is prevented from improving its circumstances. Brett describes the Marxist view of development in Uganda today:

> A new capitalist class has emerged by appropriating resources from the state and by exploiting the monopoly profits derived from irrational controls and extreme scarcity, [while] ruthless competition...devastates the lives of the losers. (Brett 1992 p20)

The geographical separation of Brett's winners and losers is highlighted by Mamdani, also a Marxist, who describes the village experience of exploitation:

> In most villages of Uganda....exploitation is mainly through forces external to the village. Internal exploitation is secondary. As a result, internal differentiation between different sections of the peasantry is not as sharp. (Mamdani 1996 p348)

Cash inflows to the Kikole economy

The marketing of crops

Cash enters the village largely through the sale of coffee to urban dealers, with much smaller amounts coming from the sale of other crops and animals. Small amounts also come to the village as remittances sent to families by relatives living elsewhere, or brought back by individuals who have spent time in an urban area, most often Kampala. The latter group are often the male heads of households who find it impossible to meet the costs of raising their family through income generated in Kikole.

Coffee berries ripen twice a year, in May, June and July, and again in October, November and December. During these periods, young men roam the village on bicycles, looking for coffee to buy. Some carry a can to use when buying by volume, and others carry a portable scale and buy by weight. All carry a large sack in which to store the coffee they have bought. At the start of the day these men are to be seen cycling through the village, an empty sack over their shoulder; in the evening those who have been successful in buying coffee return with their full sacks across the carrier of their bicycles, pushing their coffee home.

These men buy coffee either fresh or dry, the former fetching about half the price of the latter. They always pay cash. They then dry and/or store their coffee at their homes until the dealer for whom they work arrives with a truck to collect it. This dealer then transports the coffee to a factory, where it is further dried on large concrete aprons around the factory buildings, before being mechanically husked. The husks are sold as a byproduct by the factory owners, and used as a mulch on the coffee or banana plantations. Only larger, wealthier farmers can afford to buy husks and none of the participants in this study did so. The husked coffee is then transported to Kampala and exported at a price dictated by the world commodity market. Indirectly, of course, the price received by the farmer is related to the price of coffee on the global market; this price has shown dramatic variations over past years.

In the 1950s coffee processing was undertaken by farmers' co-operatives. The price paid for *robusta* coffee to farmers and their co-operatives was controlled from 1953 until 1974 by the Coffee Industry Board and later the Coffee Marketing Board (CMB), a government monopoly. Among the objectives of the CMB were two which, in practice, proved contradictory. These required it to pay a fair price to farmers, while raising money for development of other sectors of the economy through taxation. The CMB always fixed the price paid to growers below the world price so that it could achieve the second of these objectives rather than the first. After independence in 1962, taxes on exports were introduced and the prices paid to farmers slumped even further. In 1977 it reached its minimum of 15% of the export price, while the government took 66% of this price in tax. During the 1970s the price paid for coffee remained controlled, while other non-controlled prices increased dramatically, and at the end of the decade, in real terms, farmers were receiving only 10% of the price they were receiving at the beginning (Bibagamah 1996). A profit could only be made if coffee was smuggled out of the country, thus avoiding taxes (Southall 1980), and this black market (the "*magendo*" economy) grew to such an extent that in 1979/80 it accounted for over half the country's GDP (Green 1981). In spite of changes in government policies, the decline in real prices continued until the early 1990s, by which time the price of coffee had fallen by another factor of ten, to one hundredth of that existing twenty years earlier (Bibagamah 1996). These figures give an indication of the scale of the decline in wealth over their lifecourse reported by many elders in Kikole. In 1990, in meeting the requirements of the World Bank structural adjustment policies, the government removed controls on the coffee market, and export tax was finally abolished in 1992 (World Bank 1993a).

In the early 1990s, world prices rose as a result of frost damage in Brazil, reaching a peak of $4/kg, and giving the farmer up to Sh900/kg for their *robusta* coffee. At this time, the government introduced a "stabilisation" tax to remove some of the cash that was entering the economy. The World Bank (1993a) reports that in 1991 and 1992, producers' share of the international price of coffee was more than double that which they received in the mid-1980s. However, by July 1997 the impact of the Brazilian frost damage had passed and prices had fallen from their 1994 levels, and the International Coffee Organisation's guide price indicator for dry, husked *robusta* coffee was approximately Sh1500/kg.

The price received for coffee by the study participants in 1996/7 displayed dramatic seasonal variation, as shown in Table 4.1. The highest price paid for dry coffee, Sh800/kg in June 1996, was over two and half times the lowest price, Sh330/kg in January 1997. This variation was a significant factor in maintaining the relative poverty of the cash poor, who were forced to sell their coffee regardless of the price offered, compared to larger and wealthier farmers who could keep their produce until a relatively high price could be obtained. Since coffee is their sole significant cash crop, seasonal low prices could not be compensated for by the stability of or rises in the prices of other crops.

Many factors other than the world coffee price influence the price obtained for coffee by old people in the village. Among these are people's access to the market and to market information, and their vulnerability to exploitation by local buyers attempting to maximise their own profits. Beals (1975) and Hill (1972) both note, in Mexico and Nigeria respectively, the importance of attending a central market place. In such an environment individuals can appraise themselves of current market prices and negotiate with a number of buyers to ensure they obtain an optimal price for the goods they have for sale. In Kikole there were no motor vehicles available to take goods to urban markets, and it would have been necessary to hire one for this purpose. Unless one had a large amount of coffee this would have been uneconomic and none of the participants took this option, although two sold their coffee in large quantities to a dealer who collected it by truck and paid cash "at the farm gate".

Table 4.1 **Maximum and minimum prices obtained by adults aged 60+ for fresh and dry coffee between June 1996 and May 1997.** Prices shown are Ugandan Shillings (Sh) per kilogramme (Sh1000 = US$1)

Month 1996/7	Fresh coffee		Dry coffee	
	min	max	min	max
June	300	400	600	800
July	250	300	550	600
Aug			500	
Sept			500	600
Oct	250		450	500
Nov	200	250	450	500
Dec	200	200	350	400
Jan			330	350
Feb			500	600
Mar			500	
Apr			500	
May	200	250	400	500

Note: Only one transaction was recorded during months where no maximum price is shown. Fresh coffee was only sold during the cropping season, and obtained about half the price of dry coffee

All other participants sold their coffee in relatively small amounts to local buyers. These were mostly the young men described above, either unmarried and without land of their own, or married but with just a small amount of land and in need of more cash. They were provided by an urban coffee dealer with cash to use when purchasing from village people. This dealer would tell the buyer the amount he wanted to buy and the price he was prepared to pay him for it. It was therefore in the buyer's interest to pay growers as little as possible, thus maximising his own profit. At the end of the season, the dealer would also give his buyers some commission:

> We get money from people who have a lot of it and we go and find people who have got coffee to sell, and we buy it. At the end of the season I give it to the boss who had given me the money and he gives me some payment on which I have to rely because I won't get any more money until the next season. That's when starvation comes and everyone is short of money. (Young male coffee buyer)

In this situation, access to current market information was an important factor affecting a grower's ability to negotiate the best price for his or her coffee. However, there were very few radios in the village, no televisions or newspapers, and few elders were literate. Such information was therefore hard to come by. I found that market information was passed from neighbour to neighbour by word of mouth, and during the coffee ripening season my conversations with elders who lived near one of the village thoroughfares would frequently be interrupted by a passing fellow grower, usually on foot, calling out the latest prices. Similar interruptions came from buyers on bicycles who called out the price they were offering, and it was thus relatively simple to negotiate a satisfactory price. However, those who lived away from paths, or who were too frail to move around the village and meet other growers, were less likely to be in possession of the information they needed to ensure they received the correct price for their coffee.

Individuals who are not used to participation in the cash economy are particularly vulnerable to exploitation by those for whom it is a way of life (Wolf 1966, Owen 1996). In Kikole, many widows were disadvantaged because their roles as wives had not prepared them to participate in the cash economy after their husbands' deaths. Whilst some had pursued economic activities during their lives and therefore were familiar with the cash economy, those who had lived in more traditional households had, through their roles as food providers, been largely confined to the village and excluded from the cash economy. They were therefore likely to be less cognisant of the cash economy and market conditions, and vulnerable to exploitation by the more commercially aware buyers.

One morning I visited participants after they had sold coffee to a young male buyer. Two neighbours, one a relatively young and socially active older man who lived on the main path through the village, and the other a frail and very old widow whose house was hidden by trees from the path and from her neighbours, had received different prices for their coffee. On this day the man received Sh2,500 for a tin of fresh coffee, while the old woman received Sh2000. They had to accept the price on offer:

> Younger people talk with the dealers, and are often friends with them. They travel around and learn about prices in town, but the old people don't do that. They just sell to the buyer and accept what he says is the price. What other choice do they have? (Young man)

> I don't know if they are honest or not – I just have to take what they offer. (Julius Kwesige EWR=20)

The buyers were dishonest in some of their dealings, and were treated with suspicion by farmers:

> Some of them buy coffee in tins. But this is not good for the farmer. Sometimes they come with big tins, or they press the coffee in very hard and pay a poor price. (Amos Ndawula EWR=1)

Similarly, buyers could exploit the most vulnerable people, who were also the most desperate to sell, by giving them a lower price than they gave to others. There was a widespread belief that this happened but that there was nothing an old person could do about it.

> It was Christmas and everyone was selling coffee – buyers could pay what they liked. They only gave Sh3500 per tin, but there was nothing I could do and I had to accept. (Noel Katusabe EWR=9)

In Box 4.1 a coffee grower expresses his opinion of those to whom he sells his produce:

Box 4.1 Selling coffee

These young buyers don't usually use their own money, but are given some by the man they work for. They are from the village and know the people who have coffee and can exploit them by giving a very low price. They can buy it fresh or dry and take it to their own homes. Then the big buyers come and collect it from their houses with a pick-up and pay him more than the boy paid us. We can't really know the price the boy gets, but we know that they keep a lot for themselves instead of giving it to us. Buyers exploit farmers! They are the ones who know the price. They come and lie to farmers, saying that there has been a collapse in the price, but you can't wait to sell because you need the money for your problems. And you have to sell, whether he is lying or not. The buyer is the one who fixes the price.

Buyers would also offer a lower price if they suspected that, or claimed that, the quality of the coffee for sale was poor. If poor quality coffee was delivered to the factory the dealer would obtain less for his coffee, and he would then instruct his buyers that the price they were to offer growers was to fall correspondingly. Predictably, the buyers blamed the growers for producing poor coffee, while the growers blamed the buyers, saying that they failed to dry properly the fresh coffee they had bought in the village. While the growers were seriously affected by a drop in the price they obtained for their coffee, the buyers themselves did not suffer if the price fell, since their profit was based on added value rather than the absolute price so important to the grower:

> We haven't got anyone to buy our coffee yet. They say it has been spoilt by sunshine, but we want to sell it and buy sugar. (Polly Katate EWR=7)

> The coffee has been affected by the sunshine, and buyers aren't giving us a good price because they think the berries may be empty. (Ibrahim Sabiiti EWR=8)

Recently, the government has expressed concern about coffee diseases being spread through the sale, and subsequent transportation, of infected husks from coffee processing factories to uninfected plantations. Regulations have been introduced to prevent this trade. It remains to be seen whether this is successful, and if so, how it affects the price offered to growers for their coffee. Since the sale of husks adds to the coffee dealers' profits its cessation may reduce the price they offer growers for coffee.

I did not attempt to quantify household incomes precisely. Had I done so, the reluctance of individuals to provide the necessary information would have made the interpretation of this data unreliable. To give an indication of the incomes of households in Kikole, I can, however, confidently assert that few householders sold more than ten tins of dry coffee, equivalent to approximately Sh60,000 during the year of the study. The maximum amount sold was twelve sacks, each containing six tins of dry coffee, valued at about Sh750,000; the minimum was less than one tin, or less than Sh6000. The maximum would have been obtained by a large farmer selling all his coffee, after drying, to an urban dealer, on a single occasion when the price was high, while the minimum would have gone to an old woman who sold her coffee piecemeal, undried, to a local buyer.

The contribution made by other crops to the village income was usually very small, although in a few cases their contribution to individual household income was significant. *Matooke* was the only other crop exported from Kikole in any quantity, but the value of this trade was very much less than that of the trade in coffee. Small farmers said that growing bananas for sale was unprofitable, unless carried out in conjunction with another business:

> When you sell them, instead of putting the money in your pocket you have to buy grass [as a mulch], pay labourers and so on, and you don't make a profit. Those with small plots and children at home can't sell bananas as they need them to eat. You can't get rich from bananas because it costs a lot to make a banana plantation and when food is plentiful bananas lose their value, so you don't get back what you have put in. Those with good plantations have other businesses, such as being a coffee buyer. They put the coffee husks

on their plantation and get a good crop, but they don't include the costs of those husks in the costs of their bananas. (Henry Ssewanyama EWR=15)

Most *matooke* is therefore eaten at the grower's home. Any surplus is usually sold or exchanged in the village:

I sell bananas, *matooke* mostly, but sometimes *mbidde*, if any are ripe. When I need some money I send my grandson to Kilaru to find someone to buy some. The people who drink there buy them. Most of them rent rooms there, so they don't have a permanent plantation. My grandson brings them here and we negotiate a price. If we can't agree I just wait until someone else wants to buy....The man you saw today has come to buy *mbidde*, and my grandson has taken him to cut them. (George Kabongo EWR=16)

Larger farmers may sell their surplus to buyers, although this rarely occurs in Kikole. Young men with the use of bicycles, usually the same individuals as those who buy coffee in the village, also buy *matooke* and sell it in Kiguma to a middle man; he will then send it to Kampala for resale. There is a high demand for *matooke* in urban areas, and the main road from Kiguma to Kampala is busy with overloaded *matooke* trucks, frequently at the roadside with broken suspension systems. The price for a bunch of *matooke* is higher during periods of low rainfall, when they ripen only slowly. It varied between Sh800 and Sh1200 throughout most of the year, although there was a jump to Sh1500 at Christmas when demand was extreme. Immediately after Christmas the price dropped to Sh800 and remained at this low level until the end of Ramadan in February.

Mbidde were sold regularly as the prime ingredient of the local beer, *mwenge muganda*. Some growers made it themselves but the process is complex and labour intensive and if anything goes wrong, the investment is lost. Most elders sold their *mbidde* to groups of young men who made beer as an income-generating activity. Three bunches of *mbidde* sold for only Sh300. One such group is shown at work in Figure 4.1.

Sorghum, another ingredient of *mwenge muganda*, was grown, but in small quantities and by few people. One elder reported selling a tin of sorghum for Sh10,000 in June 1997. Cassava was occasionally sold. One tuber fetched Sh1000, a price which remained steady throughout the study period, reflecting the relative immunity of cassava to seasonal weather variation, and its value as a food when other foods are in short supply. Dried cassava was sold by only two people, both of whom had trouble negotiating the sale as the buyers said it was of poor quality. The price throughout the year was Sh600/kg, or Sh6000 per tin.

Dried beans are sold by the 20kg tin, and the price is very dependent on the weather, both locally and nationally. The price rose steadily from Sh2,500 per tin in June 1996 to Sh10,000 in January 1997, at the end of a poor season, when most people retained their crop for use as seed for the next season. The following season was also poor, and no beans were sold.

Two old men grew tobacco in small amounts for sale, and made only a very small additional income from this activity:

People come round here when they want some tobacco. I learned how to do this [cure tobacco] in Kampala! I was once visiting someone who was doing it as a business and I peeped into the room where they were doing it. I sell four leaves for Sh100. The whites first brought tobacco here and gave it to the *Kabaka* who gave it to the county chiefs, and they gave it to the *gomborola* chiefs, who shared it with the people. (Julius Kwesige EWR=20)

Figure 4.1   **Making *mwenge muganda***

The marketing of animals

In Kikole cattle are seen as assets to be realised only in response to a specific need, most often a cultural obligation to feed large numbers of guests at a wedding or a funeral ceremony. Other animals such as chickens, rabbits, sheep, goats and immature pigs are reared and sold to meet other costs, in particular those of taxation, education and health care. Children keep meat rabbits and sell them to raise money towards their school fees.

Adult pigs are killed and sold as meat. The high price of a mature pig is beyond the budget of most households, and so the grower has the choice of killing the animal himself and selling the meat by weight to villagers, or selling the entire animal, alive, to the man who operates the pork butchery beneath a tree in Kilaru, the village trading centre. In the first case, to maximise one's profits, one must take into account the availability of money in the village:

> We expect to kill a pig in December [when there is coffee to be sold and money available, and during the Christmas season, a time when everyone hopes to eat meat]. It is about 35kg at the moment, but people are poor these days, and if we kill it now we may not sell it all in a day. Then we will have to we keep it overnight and the blood would drain from it, making it lighter and reducing the price we could get as it will weigh less. That's why we will sell it at Christmas and then we will get money for next term's school fees. (Michael Kavuma EWR=3)

The pork butcher keeps his meat on a hook hanging from a tree and sells his meat by weight. He also cooks meat over a wood fire and sells it on skewers to the customers of the nearby village bars. He was said not to be trustworthy, and one isolated and aged widow said she was exploited by him, since he paid only about a third of the market price for her pig:

> I was going to die from poverty, so I sold a pig. They cheated me, saying it was ill, and only paid me Sh10,000. I'll use that money to buy food.....I won't get another pig because I don't have the strength to get the food for it. (Christine Nabagareka EWR=23)

Cattle require more investment than other animals kept in the village. Several young men in Kikole earned money by herding cows by day, and if one was employed, he would charge Sh1000 per cow per month. Occasional veterinary charges also had to be met. These costs could be partly offset by the farmer who kept his cows at his home and sold their milk, for which about Sh2000 per week per cow could be obtained.

Table 4.2   The values of farm animals

Animal	Value, Sh
rabbit	5-800
chicken	2-3,000
goat/sheep	30-40,000
pig	50-60,000
cow	150-250,000

Table 4.2 summarises the values of farm animals in Kikole. Since cows may fetch prices of Sh200,000 or more, sums rarely available to individual villagers, they were usually sold to urban butchers, who collected them from the village and took them to town for slaughter and sale. However, the significance of cattle ownership as insurance against future economic crisis meant that a sale occurred only rarely, and none of the cattle-owning participants sold an animal during the study period. Only when beef was required to feed guests at a village ceremony would a cow be sold in the village. In this case the price was raised co-operatively and the conditions of sale were likely to involve both cash and non-cash exchanges.

Timber products

Whole trees are sold as construction timber or for firewood. Travelling sawyers cut and saw timber on the *kibanja*. They sell their labour, while the timber remains the property of the owner of the tree. The trend towards brick- rather than mud-walled housing has created a demand for timber to fire bricks, and a single firing will consume several large trees. There are few suitable trees left in the village and only one was felled during the study period. There were, however, many *mutuba* trees, whose bark is used to make bark cloths, traditionally used for clothing and today used as blankets and to wrap bodies for burial. One elder supplemented his income by making these cloths using the bark of trees on his land. A bark cloth sells for between Sh2,000 and Sh7,000, depending on size and quality. Another man made charcoal, but this project had become a liability after he was injured falling from a tree while picking jack-fruit:

> I didn't finish the job before I fell out of that tree, so I employed someone to cut down the charcoal trees and burn them, and now I can't get a good price. Someone offered me Sh3000, but I know that the friend I made it with sold his for Sh4000. I asked him to send me someone who will buy at that price, but no-one has come yet.... I will tell people around that I have it to sell and when they hear of someone who wants charcoal they will send that person to me. I'll just keep it here until that happens. (Solomon Nsamba EWR=13)

Figures 4.2 and 4.3 illustrate two of these timber-related activities.

Bee-keeping

Two older men kept bees, although this was not always a very productive venture for one of them:

> I started keeping bees just this year. I am learning and haven't made any money yet.....I weave a hive from wood and grass and hang it in the trees. When the bees come and live in it I bring it back here to my garden and fix it to the ground with two V-shaped pieces of wood until there is honey in it. Then I make smoke with some grass and put that into the hive; the bees move to one side of the hive and I can take the combs. I leave two or three of

them as if you take them all the bees will be gone by the next day. (Noel Katusabe EWR=9 August 1996)

The bees went away! And I took a new [hive] to the bush but no bees have moved into it. (Noel Katusabe EWR=9 May 1997)

A beer bottle of honey sold in the village for Sh2,000, but the second beekeeper only sold part of his crop:

I got about six beer bottles full of honey last time....I gave some to the children when they were ill, and some to my daughter who was having a party. I only sold two bottles. (Noah Sserwadda EWR=12)

Figure 4.2   Young men bandaging the trunk of a *mutuba* tree after removing its bark

Figure 4.3   Travelling sawyers producing construction timber

Non-agricultural cash inputs to the Kikole economy

Crafts

Mats are woven by women, either for sale or for use in the home:

> We don't sell our mats. It is just that it is shameful to have nothing for visitors to sit on (Gilbert Kasibante EWR=5).

The lack of money in the village means that raw materials are difficult to obtain and buyers hard to find:

> I weave mats, but no-one buys them. I made two recently, which I wanted to sell, but eventually I gave one to a bride as a wedding gift. The materials are very expensive, and I bought these in Kampala when I went to the burial of my daughter....I was planning to sell a mat or two by the end of the school holiday so that I could send my granddaughter back to school, but that hasn't happened....The price depends on the quality; there is one called *kyaaki*, they are only about Sh2000, but *karanami* are more than Sh5000, up to Sh10,000 or even more! (Florence Mugasha EWR=11)

Remittances

Bigsten and Kayizzi-Mugerwa (1995) record that, of the 220 households in Masaka district included in their study, only 4% of household income was derived from activities outside the village, and this 4% was concentrated amongst the richer households. If they had been successful in their labour elsewhere, most often in an urban area, short-term migrants brought cash into the village on their return to Kikole. However, the difficulties attached to making a living in town made this an infrequent event. Some long-term migrants brought food, but rarely cash when they returned to visit their families, and I recorded very few instances of remittances of cash from distant relatives during the study period. This topic will be explored in Chapter 6.

Cash outflows from the Kikole economy

There were three major destinations of cash leaving the village: school fees, government taxes and health care costs. Smaller amounts were expended on the purchase of food, salt and clothing, soap, lamp oil and matches, and transportation, and some was lost through theft. Formal entertaining of visitors, and ceremonies such as weddings and funerals, could theoretically involve large payments to traders outside the village, but the poverty of those paying for these events restricted cash outflows to smaller amounts spent on beer and soft drinks for important guests. Christmas is an occasion for the expenditure of money on luxuries, but poverty prevented many participants' participation.

Education costs

The education system in Uganda   In 1964, Fallers (1964a) described the Baganda as "a people who look to education as the means to almost every desirable end, either personal or national" (p140). In 1996/7, too, many participants were making great efforts to educate their children or grandchildren, and expressed their faith in the benefits that education can provide. Traditionally, education of the most intelligent Baganda boys took place in the chiefs' palaces, where they were sent to learn the history and customs of their tribe. The object of this traditional education was the instillation of *buntubulamu* (Kajubi 1985), defined as "a very broad concept implying the possession of courtesy, compassion, good breeding, culture etc." (Murphy 1972 p44). The most favoured were educated in the *Kabaka's* palace, after which they were well placed to advance within Buganda society as adults (Fallers 1964b). Less favoured boys, and all girls received no formal education except that obtained within their household. Such education would be centred on their future roles, with boys being taught animal husbandry and building techniques, and girls learning banana cultivation, cooking and other domestic skills (Roscoe 1911).
      The first non-traditional education in Uganda was provided by missionaries, both Christian and Moslem (Fallers 1964a). In 1934, Lucy Mair found that each village had a school functioning within its church buildings, where children learned "their

prayers and their letters" (p14). She also found, however, that not all parents could afford to pay the fees demanded (Mair 1934). The government established a Department of Education in 1925, and from that date offered financial support to those who were operating schools in Uganda, but did not operate schools itself (Senteza-Kajubi 1987). In 1955, nine of the 539 schools in Buganda were government schools, 486 were affiliated with the Anglican or Roman Catholic Churches, and 46 were Moslem schools. Supervision of all primary and junior schools, however, was in the hands of the government (Fallers 1964a).

In 1970, as part of a study of young people's success or failure after leaving school, Wallace described the primary schools in three villages, situated in the Buganda, Bugisu and West Nile districts of Uganda:

> [The] schools are poor and overcrowded with minimum classroom facilities. The children are taught by rote mainly because of the lack of teaching materials and space which prevent the use of more sophisticated methods. Classes are often held in the open air, writing practice is done in the dust with a stick. The staff are often unqualified and turnover is high, especially in the private schools....Even those located within twenty miles of Kampala are often without textbooks, chalk or paper, and the situation worsens the further up-country you go. (Wallace and Weeks 1975 p10)

The decimation of the country's infrastructure that accompanied the civil wars of the 1970s and early 1980s has restricted progress in the provision of education services. The proportion of untrained primary teachers rose from 42% in 1982 to 49% in 1992, and there remains a 5% to 9% annual staff attrition rate. The majority of trained teachers are in urban areas. Only 15% of schools reported having a library in 1989, although most of these had no books, and only 49% of classrooms were classified as permanent structures (World Bank 1993b, Barton and Wamai 1994):

> The learning process is therefore reduced to the accumulation of factual knowledge through rote memorisation or copying facts from the blackboard. (World Bank 1993b p34)

In 1991/2, the government spent approximately 1.3% of GDP on education, and about 40% of this was devoted to primary education. The total spent by non-government organisations who run their own schools is more than that spent on primary education by the government. In Masaka district, government expenditure was US$9 per pupil per year, comprising US$4 for supplies and US$5 for salaries (World Bank 1993b). State schools charge additional fees to supplement government funding. No comparable figures are available for the private sector, which consists of many thousands of independent private schools, mostly run by religious organisations and other non-government organisations (NGOs). However, the fees charged by them, along with comments made by study participants, together with my own observations during visits to these schools, suggest that they have a similar low level of funding. Schools recruit their own staff. Teachers' salaries are very low, and may not be paid on time, and in order to retain their services it is necessary for schools to supplement their teachers' government salary with money raised by the Parent Teachers Association (PTA). The degree to which teachers can be supported varies between communities and between groups within communities, leading to inequalities of access to education (World Bank 1993a).

Although the government recommends that children start school at six years of age (Statistics Department 1995), education is not compulsory in Uganda. However, significant progress has been made in encouraging school attendance, for while, in 1969, 62% of Uganda's population aged five and over had never attended school, by 1990 this proportion had fallen to 37% of those aged six and over (Statistics Department 1995), and fallen further to 27% in the 1995 survey (Statistics Department 1995, 1996a). Rural attendance in 1990 was lower than urban, with the 37% non-attenders located 26% in rural and 11% in urban areas.

The education of children in Kikole Although there were 162 children aged between five and 14 years on Kikole, there were no primary schools in the village. However, a government school and Islamic, Baptist and Evangelical Christian schools were all situated in neighbouring villages, and within walking distance. The fees charged by each were similar, but both the Baptist and Evangelical schools offered reductions if the child was one of their adherents. The choice of school was made on religious and financial grounds. Most Moslem children went to the Islamic schools, while among Christians, the choice of school was usually made on the basis of cost – there was no objection raised to changing a child's religious affiliation to enable attendance at a cheaper school.

The official cost of primary education in a government school varied between Sh5,000 and Sh10,000 per term. This figure comprised a government charge of Sh1-2,000, and local charges which included a charge to register at the start of the year and a charge to sit exams at its end, a PTA fee aimed at improving the school infrastructure by repairing or constructing buildings, supplementing teachers' pay (their salaries were paid infrequently, and sometimes not at all, and they needed encouragement to stay in their posts), and to employ watchmen to protect school property at night, when armed robberies were frequent. At the Islamic school a man was employed to lead prayers at the designated times during the day.

Additional costs of sending a child to school included the provision of suitable clothes, as at some schools children were sent home if they were not wearing a clean uniform, for payment for lunch provided by the school, the costs of pencils and exercise books, and paying teachers for additional teaching which took place in the school holidays and in the period leading up to end of year exams. Pre-exam teaching required residence in the school, which involved extra fees for food and bedding. The total costs of primary education were Sh7-20,000 per term, with fees increasing as the child moved upwards through the school. Since there were three terms per year, parents and guardians were required to find between Sh21,000 and Sh60,000 per child per year.

To encourage increased attendances at primary schools, and supported by the World Bank (World Bank 1993b), the Ugandan government introduced a policy of providing Universal Primary Education (UPE) free of charge to four children per household at the start of 1997. By mid-1997 this new policy was still not fully implemented and the participants did not feel they were receiving the benefits they had hoped for. The time frame of this study did not allow a full exploration of the impact of this policy, but the comments of some village people indicate that there have been some problems with its implementation[1]:

> Now they say that every child must have a uniform! They say that since we aren't paying fees any more we have spare money and so we should buy uniforms. That will cost at least Sh10,000 per child since I have been sending them in second hand clothes I buy in Kyamulibwa. (Ibrahim Sabiiti EWR=8)

> It hasn't happened for me! President Museveni said there would be free education for four children, but the teachers want money for all of them – for the building fund, for books, for feeding the teachers... perhaps it will happen next term. (Henry Ssewanyama EWR=15)

> Museveni had pledged that he would give our children free education and that's why I sent her to school, but now they want money! Poor people's children will never learn and those who have money will have education. Should I steal another's property so that I can educate the children? (Lilliane Barenzi EWR=27)

Taxation

Tax raising and its collection was part of pre-colonial *Kiganda* life. After an annual population census had been performed, the Kabaka's agents collected bark cloths and cowrie shells from peasants. The chiefs themselves paid larger taxes of cattle or other animals, although these would have been supplied in part by the tributes of their peasant followers (Roscoe 1911, Fallers 1964c). Men were also required to labour for the Kabaka by building houses, and to contribute labour to the large Baganda army (Fallers 1964c, Richards 1964). Road mending was done by both men and women (Richards 1973a). A traditional custom of *bulungi bwansi* (lit: for the good of the community) requires individuals to provide unpaid labour on road repairs or other tasks of communal importance, and has never been formally discontinued. Mamdani (1987) reports that in 1983/4, in a Northern Ugandan village, households were required to contribute over nine hours' labour per week to communal activities such as repairs to the church or school buildings. This practice was still evident in Kikole, although to a lesser extent than described by Mamdani, and parents are expected to contribute time or money towards the construction or maintenance of school buildings.

In 1905, as part of the continuing transfer of authority from the traditional rulers to the colonial power, an annual hut tax, later replaced by a poll tax (Fallers 1964a), was imposed. Payment of this tax was intended to replace the direct services previously provided to the Kabaka (Mair 1934). At this time the Buganda chiefs expressed concern that the elderly and infirm would not be able both to pay it and meet their subsistence needs (The Regents of Buganda 1971 [first publ. 1900]). The local chiefs were required to collect the tax, and until 1926, received a proportion of the money collected as payment for their efforts (Fallers 1964a). One chief was reported granting tax exemptions on the grounds of "age or disease or some other

---

[1] Lubega (2001) reports improvements in this situation, but that schools continue to require "parents to pay for various fees as well as other scholastic materials". (p11)

physical deformity", before the First World War (Mukasa 1971 [first publ. c1925] p60). Such exemptions are still available today, and will be discussed below. Mair (1934) notes that women were not required to help their husbands in this task since they were not liable to pay tax; again, this situation remains broadly true in Uganda today. The infiltration of *Kiganda* culture by the cash economy, and by taxation in particular, can be seen in Mair's note that by 1934 a young man was said to become an adult not when he married, as was the traditional belief, but when he reached eighteen and became liable for poll-tax. She also notes that to pay these taxes he had to leave the village to earn money, indicating that the cash shortage experienced today is not purely a recent phenomenon (Mair 1934). Iliffe (1987) notes:

> Taxation drained money from rural communities and contributed to that "cash hunger" so widespread in the Third World. (p154)

In pre-colonial times the payment of taxes in labour rather than the usual shells or bark cloths was allowed, and a month's labour was deemed of equivalent value (Wrigley 1964). In 1934, and again 1950, the amount of tax to be paid by an unskilled labourer approximated one month's cash income (Mair 1934, Richards 1973a). This system had developed during a period when individual incomes were relatively similar, but its inequity grew as economic differentiation resulted in wide variation in individual income. In 1953, the Wallis report argued that a flat rate of taxation was no longer appropriate (Wallis 1953, cited in Davey 1974). A graduated tax based on "the value of visible possessions" (Davey 1974 p30) was introduced in Bukedi district in 1960, but perceptions of its inequity resulted in riots. As a result a new basis upon which to calculate individual liability to graduated tax was introduced:

> It is clearly undesirable that assets which are not revenue earning, eg houses used by tax payers for their own occupation – should be taxed. But it is equally clearly desirable that assets which are actually or potentially income earning, for example cattle, should be taken into account in assessing tax. (Ministry of Local Government 1960, cited in Davey 1974 p39)

This remains broadly the situation today: those in formal employment are liable to pay income tax, while all others are required to pay "graduated tax":

> Graduated tax is a crude form of income tax levied in Uganda upon the entire population of able-bodied adult males and some women by the district administrations and urban authorities where they reside. (Davey 1974 p31)

In Kikole today, tax assessment is made and taxes collected by the local *gomborola* chief:

> First we enumerate – we find everyone who is old enough to pay tax. That is from eighteen up to death. Men whose wives have died have to pay as they always have done, but only women who earn money pay tax. A woman with a lot of land doesn't pay tax these days. But some old women have nothing, anyway! You pay Sh750 for each chicken

you have, goats are Sh3000 each, and sheep the same. More for a cow. Everything is counted, including crops; coffee, beans everything. We write on the form how many acres of coffee or bananas, and how many animals and add it all up. And if he has a business, say hair cutting, or laundry, or makes beer from *kisubi,* sells pineapples, has a factory and so on, then there is more tax for that. Everyone gets a G-form, which tells them how much they have to pay and how it is calculated. The minimum amount to be paid is Sh13,000, which is demanded from adult men who do not own land, nor have another income, and the maximum of a farmer is Sh88,000. A man can appeal against his assessment and then the court comes to his *kibanja* and checks that the parish chief has been fair. (Tax collector)

Perhaps unsurprisingly, some participants did not agree with their tax assessments. Specifically, they claimed that no account was taken of the declining productivity of their land as they aged and became less able to cultivate it:

My tax is assessed on the amount of bananas and coffee I have on my land. They make no allowance for most of it being fallow because I can't dig it any more. I was assessed to pay Sh16,000, plus Sh2,000 for the development of schools.

They come and look around, and decide what I should pay. They don't survey all my land, but give me a form saying how much I have to pay. I have no choice – you cannot refuse. They are people from this village, and they sit at their headquarters and decide how much they want each person to pay.

I used to have a big income, but not any more. Now I am overcharged in graduated tax by the government because I still pay as though I am very rich and if I have five tins of coffee all of it will go in taxes. They come and look at your coffee, but they use their records to charge you.

In spite of their grievances, during the study period none of the participants appealed against their assessment. Tax exemptions are available to the frail or disabled, but had to be requested:

The Commissioner of Taxes comes here once a year and on that day old men come and speak to him: "I have been paying my taxes for 60 or 70 years and now I am weak and can't pay my tax, and I am poor. I can only just get enough to feed myself, that's all." Sometimes, when you bring these old people here they stand silent, and when the Commissioner asks them questions about their circumstances they don't speak. They are scared because they aren't educated. Often the Commissioner says to a man that can't speak for himself "Go! You will pay taxes!", so we speak for them. Or he may give either a permanent or a temporary exemption. If it's a temporary one they will check again next year, in case he has become rich again. This is a free service – there is no charge for a tax exemption. (Tax collector)

Box 4.2 contains a description of one participant's experience of applying for a tax exemption, and indicates that although there were no formal charges, informal payments were made:

Box 4.2 Obtaining a tax exemption

I tried to get an exemption once. I expected it last year, since the year before the District Commissioner had promised me I would get one. Up till then I was assessed at Sh12,000, but he reduced it to Sh9,000. But then last year he was replaced by another man. We were all invited to go to the *gomborola* chief, to ask for an exemption, but we didn't meet him face to face. The parish chief introduced each of us with our problems to the district commissioner, and asked for an exemption. Just at that moment the *gomborola* chief came out of his room and said that the parish chief had been bribed by us, and so the commissioner refused to exempt us. Then they assessed me back to Sh12,000 and later put it up to Sh18,000! I had to borrow to pay that and haven't repaid it yet....We had bribed that man, of course – he won't do anything for you if there's no profit for him, will he? He charges the equivalent of a cock, but I think that time the *gomborola* chief was angry because he hadn't had a share, and decided to make it public. This happens all the time – many of us have had this experience.

In Kikole the tax collectors make every effort to collect the monies during the cropping season, when villagers have money in their pockets, but the villagers often have other priorities for their money, particularly if they have debts to repay to neighbours. The collectors were patient, but if tax is not paid then one eventually becomes liable to imprisonment. The breadth of the assessors' powers were also seen in their attitude to imprisoning defaulters:

> We don't take them to court where the magistrate will charge them, fine them, and imprison them and then, after that, they will still have to pay their taxes. Instead we bring them here to the *gomborola* prison just to punish them a bit. We keep them here a day or two and them send them home again with instructions to come back with the money. We give them a choice – prison or pay the tax, and they always choose to pay! If we did nothing they would tell their friends and no-one would ever pay! (Tax collector)

Some participants accepted this risk of imprisonment:

> I haven't paid my taxes this year and they came around arresting people who haven't paid. I said they should take me as I don't have the money – so they left.

After paying tax one is given a receipt which one is obliged to carry at all times. They are usually wrapped in an old plastic bag, and tucked into the waistband of their owner's trousers. Some participants let me examine their papers, which had clearly been slowly disintegrating for years, and in some cases, decades. Tax receipts have to be shown to inspectors on request. These men set up road blocks and arrest people who cannot prove they have paid their tax (Davey 1974, Robertson 1978). I had direct experience of this while on a *matatu* [a public minibus/taxi] to Kampala, when tax inspectors stopped the vehicle and demanded to see all the men's tax receipts. One man who was unable to do so was arrested and removed from the bus.

Graduated tax was introduced in Buganda in 1954 and has remained in force since that date. Only one of the women participants, the midwife mentioned above, was required to pay this tax, and several of the men had obtained exemptions on the grounds of age or infirmity. In 1996/7, a typical sum paid by a small farmer in Kikole was Sh18,000, which I estimate as between 30% and 50% of the income they obtain through the sale of coffee. At the extremes however, this figure may well be as high as 70% for the poorest who are required to pay at least Sh13,000, and at least as low as 15% for the largest producers for whom the maximum graduated tax is Sh88,000. The inequity in the application of graduated tax in Buganda, cited by Mamdani (1996), is visible in Kikole.

Health care costs

Uganda's health indicators such as infant mortality and life expectancy at birth are among the world's worst (World Bank 1993b). It was therefore likely that study participants, doubly disadvantaged through age and rural residence, would face health problems and need to respond to them. For all but the most basic care, it was necessary to leave the village and consult practitioners who charged for their services. Charges ranged from Sh600 for a few tablets to Sh40,000 for inpatient hospital treatment, and are set out in Table 4.3 below. The availability and quality of health services and the issues surrounding their use by study participants will be discussed in the following chapter.

Other cash outflows

Weddings, burials and funeral ceremonies are, ideally, lavish occasions, demanding the expenditure of large sums of money on food and drink, much of which would be purchased from outside the village. A number of specialist businesses aimed to meet the needs of those holding these events. For example, a man with an electronic megaphone could be engaged to cycle through the village announcing a death and the time of the burial, or one could buy the services of ceremonial decorators who would erect stages and shelters and decorate them with colourful fabrics and tinsel. For a wedding ceremony, furniture had to be hired, and since music was a necessity, a "ghetto-blaster" with battery and tapes was also required. A car would be needed to transport the bride and the guests of honour to a wedding, while the priest charged to attend and officiate. None of these services was available within the village, and substantial sums of money left the village economy as a result of their purchase.

Many individuals try to make a living through trading with village people. Within the village there were three *dduukas* [small shops] selling goods such as packaged foods, cosmetics and medicines. Since they were residents, the profits made by these shopkeepers remained in the village, but their costs, paid by them to wholesalers in towns, were lost to the village economy. Fallers (1964a) reports that in the early 1950s these small shops were frequently unprofitable. They mostly dealt in low profit-margin goods such as sugar, cigarettes and kerosene, and their owners were, through lack of contacts or experience, unable to deal in more profitable goods such as clothing and hardware. My observations were that this is still very much the case, although I would suggest that the limitations on goods offered for sale are a result more of the lack of cash, and therefore of custom, in the village rather than in the difficulties listed by Fallers. In addition to these *dduukas* there were many informal shops in operation, with a household investing a small amount of surplus cash in a very limited stock, perhaps a single item, with a view to making a profit on its sale. For example, one household was found at one time to be selling half kilogramme bags of salt at Sh300 each, while another sold *mukene* in small quantities for as little as Sh100. Other visiting traders toured the village on bicycles, selling fresh meat and fish, and some villagers purchased goods at one of the several small weekly markets in nearby villages.

Table 4.3   Reported costs of a course of treatment at all health care providers consulted by members of participants' households, July 1996 - June 1997

Health facility	Reported costs per course of treatment (Uganda shillings)
Vila Maria Hospital (inpatient)	15,000 - 40,000
Ssematiko's clinic (inpatient)	10,000 - 21,000
Ssematiko's clinic (outpatient)	600 - 10,000
Government dispensaries	2,300 - 10,000
Local midwife's clinic	1,200 - 6,000
Kawanga (independent biomedical practitioner)	3,000 - 4,000
Kagimu (independent biomedical practitioner)	1,000 - 11,000
Dr Kawooza (independent, qualified biomedical practitioner)	5,000
Dentist	4,000 - 6,000
Bonesetter	6,000
Self treatment	500 - 1500

Theft was frequently reported in Kikole. Since the village was short of cash, material goods were the usual target of thieves:

I have been robbed twice – the second time they stole drugs and mattresses from me. (Emma Kabenge - midwife EWR=4)

Recently people came and stole from me when I wasn't around. I had a small shop, but everything was stolen, and I need money for clothes, food, meat and fish... (George Kabongo EWR=16)

Thieves took a whole basket of dry coffee! It isn't right that an old woman like me should have her property stolen. It was taken by my daughter and son-in-law, while I was picking more coffee up by the road with my granddaughter. When we came back, someone had been into the house and taken the basket. We looked around and found it down there, under that coffee tree. It was empty. (Edith Manube EWR=29)

Cash was occasionally sent, as a gift, to people living outside the village. Most often, these were small contributions towards the costs of the wedding or the burial of a family member. Larger transfers of cash to destinations outside the village were made by a few, relatively wealthy, participants. Two women, for example, regularly paid secondary school fees for a number of grandchildren living in Kampala and Masaka.

The circulation of cash within the village

In Kikole today, just as Lucy Mair noted over 60 years ago (Mair 1934), once the villagers are in possession of their seasonal income it is rapidly consumed in the payment of debts, school fees, taxes, and household essentials, and only a little remains to circulate in the village. One of the more visible manifestations of this circulation is in the trade in beer at the village bars. The area where the bars were located was the centre of social and economic activity in the village and the quality of the experience to be obtained there can be gauged from its local nickname: "*Kilaru*" ["The Madhouse"]. Local beer was, as mentioned above, made in the village with local ingredients. It was then taken by the bar owner, by purchase outright or on a sale-or-return basis, and sold to the village people:

I made some beer today! I have taken two jerry cans to a bar in Kilaru and they will pay me after they have sold it. They will sell a jerry can of beer for Sh3000, and give me Sh2000. The rest is their profit. (Vincent Katorogo EWR=18)

The takings, after the deduction of costs and a profit margin were then reinvested in the purchase of more beer. Beer making was a widespread and obviously enjoyable activity, and commonly seen and heard in Kikole, just as was the unsteady gait and garrulous greeting of a drinker returning home from Kilaru.

Several households, taking advantage of their superior roadside location offered "fast food" such as fruit juice, pancakes, sweet bananas and passion fruit to passers-by, while some individuals, usually children, walked the village paths selling these items. Cash also circulated within the economy as the contributions made by each household towards the costs of one of their number facing the relatively enormous costs of staging a ceremony, whether a burial, or funeral, or wedding. This is in effect a revolving fund, largely separate from the rest of the village economy; contributions cannot be used to meet other expenses, and nor can the credit one has established be called upon unless one has a ceremony to pay for. It is hard to estimate the size of this fund, but poor people's contributions were only of the order of a few hundred shillings, and so it is unlikely to amount to more than, say, Sh50,000. This fund and the exchanges that take place within it are explored further in Chapter 6.

Small amounts of money circulated in the village economy as payments for the supply of labour. The shortage of cash in the village severely restricts the opportunities available to someone whose land does not produce enough cash for them to meet their household needs. Family members, friends and neighbours are usually equally short of cash, and breeding and selling livestock, or engaging in craft activities requires capital investment without the guarantee that a customer will be found. Wage labour, when it can be found, is an option open to someone without capital, or unwilling to risk their capital, and with the requisite physical ability.

Bigsten and Kayizzi-Mugerwa (1995), working in Masaka district, noted that wage labour was not a major part of household economies in 1990, and this remains the situation in Kikole today. The cessation of the supply of migrant labour in the 1970s (Jamal 1991) changed the wage labour market dramatically:

> There used to be people who came from Rwanda and Burundi to work. They would work the whole season without pay, and then get paid after the crop had been sold. Then it was possible to farm coffee on a large scale, but after they stopped coming most of our coffee went fallow as there was no-one to dig the land. These days, if someone works for you he wants money on the spot, each day, which is not possible. (Ibrahim Sabiiti EWR=8)

Consequently there was no formal employment available in Kikole, and casual wage labour formed a relatively small component of the village economy. It was, however, crucial to the well-being of the households of the few who undertook it. Wages were low, and only two of the poorer, younger elders, one man and one woman, regularly worked for others:

> When the coffee harvest is good I have money, but if it's not I go and work for other people to get money for salt and for food. At the moment I am cutting grass for someone. He is paying me Sh2000, and it will take me six days to finish the job. He has given me Sh500 as an advance and I will get the rest when it is finished. Then I have another job to do – building a small house. That is for Sh5000, and will be about 12 days' work altogether. (Solomon Nsamba EWR=13)

> I do any type of work in the village to pay school fees for my grandchildren. At the moment I am trying to raise Sh10,000 that we have to pay for the school buildings and feeding the teachers, so although I am not well I am having to work for a neighbour. I will get Sh3000 for this job, and it will take me eight or ten days to finish it, as I am very weak. (Rebecca Nambiru EWR=22)

Apart from these two who regularly engaged in wage labour, I found one other example of a man and wife who did just one paid job during the study period:

My wife and I are working on a small job for Sh4000 in the village at the moment. After that, I will plant my seeds if it has started raining. What will we eat if I work in the village but do nothing at home? (Henry Ssewanyama EWR=15)

Not all money spent on health care left the village, as it was fortunate in having a retired biomedically trained midwife as a resident. This woman, one of the study participants, provided midwifery and basic health care services for a fee, although frequently on credit, which often was not repaid. There were also a traditional midwife and a spiritual healer resident in the village, who also charged for their services. Both extended credit to their customers and were prepared to accept payment in cash or kind. The informal and individual nature of these arrangements makes it impossible to estimate their magnitude or frequency.

Credit

Scott (1976) reports that Burmese peasants prized tenancy above wage labour, and land ownership above tenancy because of the increased access to credit these afforded. A land owner could use his land as security against a loan, while a tenant could rightfully expect his landlord to advance him money, whereas a wage labourer had very little opportunities to borrow. In Kikole, as we have seen, there were no landlords for *kibanja* holders to turn to, and although the *kibanja* itself did have a limited commercial value, its importance to the maintenance of household well-being largely prevented its use as security against a loan. Most women, since they did not own land, did not have even this option open to them.

Although the importance of access to credit to the promotion of household well-being is well recorded, particularly for women and female-headed households (Todaro 1989, Gittinger 1990, World Bank 1993a), the villagers of Kikole have no institutional access to credit. Nevertheless, the seasonality of cash income in Kikole makes credit an important factor in the lives of both rich and poor. Many individuals were obliged to go into debt between the coffee seasons, and shopkeepers, running a business at a time when very few people have any money to spend, were obliged to make credit available in order to attract customers. Many participants ran up debts of a few thousand shillings with their local shopkeeper, or a neighbour, friend or relative. Larger advances, such as for the payment of school fees or health care costs, required more formal arrangements to be entered into, involving written contracts drawn up and signed by both parties, and witnessed by a third party. Debtor-creditor relationships are described in Chapter 6.

Cash reserves

One of the most acrimonious family disputes which took place during my time in Kikole erupted when the cash reserves of the head of a family, hidden in the roof of his house, were lost when the building was destroyed by fire. Mair (1934) noted that most households held such a reserve for emergency use. It is my belief that most households in Kikole today also hold some cash in hand to help cope with unexpected problems such as a sudden illness or a death, although they are reluctant to discuss the amounts involved. Local insecurity, and a consequent fear of burglary, is the likely cause of this reluctance.

The non-cash economy

In view of the cash shortage described above, it should not be surprising that non-cash exchanges are common in Kikole. Exchanges in kind are frequent, the most often seen being the exchange of food for labour. Married women, for example, who have the responsibility of providing their household with food, but who are frequently excluded from the cash economy, may not always have enough food from their own land and will work on another's in exchange for a bunch of *matooke*, or some cassava. Rebecca, a grandmother caring for grandchildren, whose wage labour was described above, also worked, with friends, for food when her own small plot of land was unable to feed her household:

> First we go to the person who has work for us and we survey the job and decide on the amount of cassava we want for doing it. If they accept we start digging. When we have finished we share the food equally amongst ourselves. If the person doesn't have enough food for us we leave the job and return to our own fields. (Rebecca Nambiru EWR=22)

Although wage labour was only available from the wealthier farmers, poorer elders needed to employ labour as their own physical abilities declined. One of the poorest was unable to do much physical work herself, and used some of her very limited food supply to employ labour:

> I was able to employ porters as I had some ripe bananas at the time. People move around the village looking for work – women or young men, not married men; it is the wives that do it – they ask if you have food for digging and then you negotiate. (Lilliane Barenzi EWR=27)

Scott (1976), notes the importance of reciprocal exchanges in the moral economy of the peasant. Their primary objective is not necessarily the maximisation of personal gain, and therefore they do not always involve exchanges of goods or services of equal monetary value. As such, they have been excluded from this exploration of the economy of Kikole, and are explored in detail in Chapter 6.

Non-cash reserves

Whilst a few of the wealthier households had stores of coffee, waiting for sale when a high price was obtainable, and numbers of cattle that could be sold when cash was required, most had very limited non-cash reserves of any kind. Smaller animals were the most frequently owned, and most easily liquidated material assets. The poorest considered their furniture, their agricultural implements and even the iron sheets, doors and windows of their houses as realisable assets. This was visible in the relationship between house construction and relative wealth as shown in Table 4.4. Those who lived in cheaper, less durable mud-walled or grass roofed homes were poorer than those who lived in brick houses.

Table 4.4   Materials used in house construction and mean emic wealth rank of heads of households aged 60+

Wall material	Roof material	Mean emic wealth rank	
Brick	Iron	7.3	(n=6*)
Mud	Mixed	12.5	(n=2)
Mud	Iron	13.6	(n=13)
Mud	Grass	26.0	(n=5)

* = Number of households

Summary

In Kikole, almost all cash income is generated by agricultural production. However, most of this income then leaves the village and therefore does not contribute to its development. Firstly, all adult males have to pay government taxes. Secondly, fees are paid to urban-based institutions, such the providers of educational and health care services. Thirdly, since coffee growers were unable to participate on equal terms with buyers, who had greater knowledge of market conditions, they received less than the optimum price for their produce, and were effectively relieved of a proportion of the added value created by their labour. Thus the quality of participation in the coffee market limited the cash that came into the village. Cash that did so was largely required to meet the various unavoidable costs demanded of all village people, by external agencies. These costs were legion, and included, apart from those major expenditures listed above, smaller payments for transportation, fees paid to the church, bribes to local politicians, and the costs of household goods and clothing, and cultural events such as weddings and funerals. The last of these is becoming increasingly significant as a result of the ongoing HIV/AIDS epidemic and will be discussed further in Chapter 7. In this way the population is rendered cash-poor, with the result that there was virtually no market for the other major "crop" produced by Kikole farmers; that is the labour of their adult children, who were then forced to leave the village and look elsewhere for work.

When problems that require the expenditure of cash arise, such as the need to pay school fees or health care costs, the primary concern of many elders is to meet their immediate cash needs, rather than to maximise their cash income. It would, therefore, be misleading to present all the cash and non-cash exchanges described in this chapter as purely profit maximising transactions. Some, such as the sale of a large quantity of coffee to a single dealer, were undoubtedly between parties who were both equally intent on maximising their profit from the transaction. Others, such as those between aged and isolated people and younger, better informed coffee buyers, were very different. The values which mediated the exchanges, and which varied between individuals and between the participants in the exchange, were not necessarily those of profit maximisation, and it is at this point that a neo-classical or Marxian analysis of the village economy is seen to have dubious utility. There is a need to set these transactions within a different conceptual framework, one which does not see profit maximisation as governing all economic behaviour. This framework will be developed in Chapter 6.

This chapter and the preceding chapter have described the elders' use of land and labour in the maintenance of well-being. Poor health status limits individual capacity to exploit these resources since, without physical strength one cannot dig one's land or labour for others. Elders experience an inevitable decline in their physical ability, and therefore in the degree to which they can meet their needs through their own labours. In the following chapter I address the factors that influence the health of the aged in Kikole, and explore their use of the available health care services.

Chapter 5

The Health of Elders in Kikole

Things are bad these days. Old people die simply because they are poor and can't pay hospital bills. (Emma Kabenge EWR=2)

Elders are vulnerable to the same infectious or environmentally produced diseases as the rest of the population, and when considering their health problems it is difficult to differentiate between illness related to environment or lifestyle and that which is age-related (Nayar 1996). Consequently, whilst advancing age may produce a decline in physical and/or mental ability (Kane *et al.* 1994), there is no justification for assuming that ill health is an inevitable concomitant of the ageing process. In considering the health of Kikole's elders it is necessary, therefore, to identify and understand relevant environmental or lifestyle factors, and to differentiate the problems they cause from those that are associated with advancing age.

In sub-Saharan Africa, available data indicates that among those aged 65 years and over, circulatory illness causes 41% of deaths, followed by infectious and parasitic illness (19%), cancers (9%), injury (2%), with 28% being from other causes (Feachem *et al.* 1991). No quantitative data are available to describe the health status of the aged in contemporary Uganda as a whole, or to describe the main causes of their deaths. However, some revealing data is available from 1971, of morbidity among a self-referred (and therefore unrepresentative) group attenders aged over 50 at a peri-urban clinic (Kakande *et al.* 1972). Among this group of 49 men and 68 women, 53% of men and 65% of women were hypertensive, 79% of men and 44% of women were anaemic, and 29% of the total had evidence of heart failure. Sixty eight per cent had hookworm infection and, after anthropometric assessment, 42% of men and 34% of women were classified as wasted. These figures should be considered within the context of the several years' political unrest in Uganda prior to 1971, which may have influenced the health of elders at that time.

The health problems faced by older adults are frequently the results of earlier life experience. Years of continuous hard labour, of a poor diet, or of having given birth to and reared many children may all lead to health problems in later life (Tout 1989). I shall show that, in Kikole, environmental conditions are poor and access to food and to health services is far from adequate. Here morbidity, in the form of infectious disease and disability, compromises the well-being, and possibly the survival of elders, through reducing their ability to labour and to grow food:

We work too hard. I work from morning to midday and it is bad for me. And we eat badly; we have food, but we eat it without sauce. (William Muwonge EWR=4)

We old people are going to die from many causes. Like eating badly and not sleeping well. Imagine sleeping without any good cover to keep you warm – how can you stay healthy? (Vincent Katorogo EWR=18)

Of the 30 elders in Kikole only one was obviously physically disabled. This was a man with right-sided weakness, fully alert and orientated but able only to walk short distances, who was cared for by his wife and daughters. Three old women of varying degrees of frailty were dependent on others to meet most of their needs. All other participants were more or less independent, and caring for themselves. As I shall demonstrate, however, this does not necessarily mean that their health needs were met in full, or that they were in good health.

Biomedical health services in Uganda

In 1970 Uganda had one of the most advanced health care delivery systems in Africa, with a well integrated network of rural and urban health units and large regional hospitals, well staffed with trained health workers at all levels (Dodge and Wiebe 1985). No charge was made for treatment and health facilities were well attended (Whyte 1991). At this time 75% to 80% of health expenditure was directed towards curative rather than preventative services (Scheyer and Dunlop 1985).

During the unrest and insecurity of the late 1960s, 1970s and early 1980s these services experienced severe difficulties. Over half the nation's doctors and many other health workers either left the country or were killed. Hospital building and maintenance programmes were halted, and there were severe shortages of drugs and other essential supplies (Scheyer and Dunlop 1985). In this period, with the increase in the incidence of communicable and vector-borne diseases, especially tuberculosis, measles and malaria, and, later, HIV infection, Uganda's position in the African nations' population health status "league" declined from first to fortieth (Macrae *et al.* 1996).

Missionaries had been providing health care in Uganda for many years, and their contribution became more significant as the government health care system declined. Non-government organisations (NGOs) and private practitioners also appeared in large numbers, many of whom had previously been employed in the state system. In response to the absence of the free treatment that was previously offered by the government, many poor people turned to self-medication (Whyte 1991, Macrae *et al.* 1996).

In 1981, with the arrival of a degree of political stability, international agencies began to support the rehabilitation of the government health services (Dodge 1987, Okuonze and Macrae 1995), a task which was largely complete by 1991. In government facilities, patients are now required to pay for their drugs, and they continue to have to pay non-government providers for their services. The emphasis on curative rather than preventative services persists, however, with 90% of health service funds being expended on the former (World Bank 1993b).

Health service utilisation has been shown to decrease with the distance it is necessary to travel to obtain treatment (Good 1987). In 1991, the numbers of functioning government health facilities had grown sufficiently for the World Bank to state that proximity to a health facility is not generally a problem in Uganda (World Bank 1993b). In 1992 the Ugandan government estimated that 27% of the population lived within 5km of the nearest government health unit, and 57% within 10km (Okello *et al.* 1998).

The availability of and access to a service does not, however, ensure its utilisation, if the service offered is not acceptable to the population it is intended to serve (Colson 1972). Gershenberg (1972) found that, in Uganda, attendances at hospitals and outpatient dispensaries declined by 50% for every two miles (3.2 kilometres) of travelling required. The World Bank has reported that nationally the government provides 40% of modern curative outpatient services, with 35% being provided by private practitioners and 25% by NGOs (World Bank 1993b). A 1991 Makerere University (Kampala) study showed that in Masaka district, while 35% of 191 representative households named a government facility as their nearest health service provider, only 5% named this facility as their preferred provider, raising questions about the quality of service provided by the government facilities (World Bank 1993b).

Environmental threats to the health of aged people in Kikole

A health transition accompanies economic development, which consists broadly in a decrease in the significance of infectious and parasitic diseases and a corresponding increase in non-communicable, degenerative or environmentally induced illnesses (Omran 1971, Caldwell 1993). Thus the conditions of most concern to older people in developed countries today are "lifestyle" conditions such as rheumatic and cardiovascular problems, and cancers (Olshansky and Ault 1986, Parkin *et al.* 1993). In sub-Saharan Africa, however, communicable diseases remain the main causes of death (Feachem *et al.* 1991), and transmission of these diseases is greatly facilitated by poor environmental conditions, particularly poor housing conditions and poor sanitation (Kloos 1994).

Housing

Poor housing has been associated with an increased incidence of infectious diseases, including tuberculosis (Last 1992). Tuberculosis is a major cause of death in developing countries, and increasingly so since the arrival of the HIV/AIDS epidemic (World Health Organisation 1989, de Cock *et al.* 1993).

The materials of house construction in Kikole were described in Table 4.4. A new brick house cost a lot of money, and therefore could only be built by the wealthier villagers. Only one elder (Ibrahim Sabiiti EWR=8) started to build a new brick house during the study. He reused his existing iron sheets on the new roof, and estimated that the house would cost him Sh130,000 to build. Mud houses were cheaper, as their materials were readily available. The wooden poles to construct the framework, and the mud to cover the walls usually came from one's land. The reeds to act as the wattles and the grass to thatch the roof came for the cost of cutting and transporting them from the local swamp. Mud and grass houses, in particular, need regular maintenance as the weather and termites take a heavy toll on them. Elders are more vulnerable to illness or accident if their home is not in good condition, but have trouble carrying out this maintenance, or paying someone else to do it:

> My husband built a mud house long ago, behind where this one now stands. Around the time of his death it was eaten by termites and fell down. My daughter, who now lives at the lake shore, sold some of the iron sheets from that one to pay for this one to be built. But now this one is being eaten and will fall down soon. My son should help me but he has never given me even a piece of cloth! (Lilliane Barenzi EWR=27)

> Ants are eating the poles of my house, and it will fall down soon. I tried to remove the queen from the anthill behind the house but they are rebuilding it now. (Solomon Nsamba EWR=13)

Figure 5.1   The house of a coffee buyer

All mud walled houses had earth floors, as did three of the six brick houses. The other three brick houses had concrete floors. Mud walls and earth floors provided both shelter and breeding sites for insect pests. Generally, wealthier households had fewer problems with insects:

> We don't have a problem with pests because we spray the house and wash our clothes. (Amos Ndawula; Brick walls concrete floor, EWR=1)

Fleas are common in this village because people leave dry grass on their floors. But the grass you can see here doesn't stay long, and [my daughter] will bring some fresh soon. She will pour boiling water on the floor before they put it down and the fleas which were being born in the old grass get killed. (Gilbert Kasibante; Brick walls concrete floor EWR=5)

Figure 5.2   A newly constructed mud walled, grass roofed home

Poorer households, with poorer quality housing, experienced more insect problems:

We have a lot of fleas and maggots living here, because my son Joseph used to sleep in the living room and they bit him. He has gone now, but they don't die when there's no-one sleeping there – they just hide in the dust. I poured boiling water on them once, but they were back again within a month. I dig bedbugs out of the wall with a stick, and kill them, but the fleas just bite us until they are tired.... (Rebecca Nambiru; Brick walls (very decayed) earth floor EWR=22)

I have bedbugs and fleas in my house, but I have no money for pesticides, so they bite me when they want to. (Juliet Kisaakye; mud walls and floor EWR=25)

While cooking was done outside, stores of food were kept secure since they were liable to be stolen, or eaten by animals. Some households had no secure storage space outside their houses and rats came into the houses at night, looking for food:

I have rats inside my house and they give a lot of trouble. See how they've been eating my feet? (Solomon Nsamba EWR=13)

Houses are constructed without fireplaces or chimneys, and cooking is usually done in a separate kitchen, constructed very nearby. Some people did not have a kitchen, and usually cooked outside their houses, but had to move inside if the weather was bad:

> We have no paraffin. When the moon is bright we eat outside, but when it's not we eat inside, by the fire. Recently the rain has been bad and we have had to cook and eat by the fire inside the house. (Rebecca Nambiru EWR=22)

Several participants cooked over a fire inside their house, the smoke of which presented health problems. One frail woman, whose house had a particularly high front doorstep, preferred to cook inside her house:

> The problem is that I don't have the strength to carry pots from outside and I am afraid I will fall over. I used to have it outside, but it was hard to get in and out of the door at night, and once I fell over. So now I have it in here, but I still fall down sometimes. When I cook near my bed I can cook lying down and that is easier. (Irene Mutondo EWR=30)

This aged woman complained of both sore eyes and a cough. Sore eyes may also be due to poor lighting:

> Everything is very expensive! I don't use paraffin, but have a fire in the house at night and use the light from that. (Christine Nabagareka EWR=23)

Many elders care for one or more of their grandchildren. These additional household members can cause overcrowding and thereby create health problems:

> I will have to accept these orphans – where else would they go? But I need a bigger house as I don't have room for them now. I have a few bricks here so I might just build a store outside and make more room in the house. (Grace Nanteza EWR=6)

Some had beds, but most slept on mats spread on the ground. As Figure 5.3 shows, pests were a problem. Although temperatures rarely fell below fifteen degrees centigrade, many felt the cold at night:

> Since you have come here to learn the problems of old people I want to tell you that a big problem is the lack of something to cover ourselves with at night. Even if you eat badly you can still be OK if you have a blanket. (Henry Ssewanyama EWR=15)

A few of the wealthier households had a complete set of bedding, but the poorest had little:

> I sleep on a bark cloth on the ground and I cover myself with my dress. I'd sleep better if I had a blanket! (Juliet Kisaakye EWR=25)

Figure 5.3   Removing insect pests from bedding

Water supply

Bradley's classification of water-related diseases (White *et al.* 1972), which was developed in an East African context, and has since proved to have global applicability, shows the several routes by which infectious or parasitic organisms may be transmitted from one person to another. Those unable to access sufficient, good quality water, are shown to be at higher risk of infectious diseases than others (Cairncross and Feachem 1993).

In Kikole, apart from uses associated with food preparation and domestic hygiene, water was needed to make bricks, to make and repair mud houses, in the preparation of beer, and to water animals. The village had two protected springs, where concrete aprons surrounded a water outlet. These flowed continuously and no pumping was required. Surplus water did not drain away well, and it was necessary to stand in about 15cm of dirty water to place a jerry can under the outlet. They were located in the valley below the village, at the bottom of a steep hill. During wet periods the path became muddy and difficult to negotiate. A third source of water, used by the few houses nearby was located higher up the valley. This water source was an unprotected well, and water was collected by dipping a vessel into a murky pool. The most distant houses in the village were situated about one and a half kilometres from their water source, and 100 metres above it. For these households obtaining water was a problem:

If I need water I have to go down to John's house and ask him to get some for me...or I beg for it from my relatives down there. (Christine Nabagareka EWR=23)

Rainwater was commonly collected from roofs, particularly by those living furthest from the springs. A split banana stem served as a gutter, and a coiled banana leaf as a funnel to direct the water into a jerry can. A heavy shower would produce a flurry of activity in these compounds, as some household members rushed to gather in coffee and other crops drying on the ground, while others erected equipment to collect the rain water. It was possible, if rain fell every day, for a small household to survive, but no more, without collecting water from the spring.

Ownership of an appropriate vessel in which to contain water, and the physical ability to carry this vessel from the spring to the home were essential to the maintenance of an adequate household water supply. Jerry cans of various sizes were employed, and all households had at least one of these. One, however, did not have one in good condition:

My grandson collects it, but the jerry can leaks and only stays half full. (Juliet Kisaakye EWR=25)

Water was usually collected from the spring by young children. In households with no children, or with only very young children, adults collected water:

I take these two small jerry cans, and when my wife has time she takes the bigger one. That one is 20 litres, but it is too heavy for me to carry from down in the valley. (Solomon Nsamba EWR=13)

If no support was available from family or neighbours it was necessary to pay someone to bring water. A lone woman, with no children living nearby, paid a man to get her water:

If my grandchildren are here they get it for me, but if not I pay a man Sh100 per jerry can to bring it here. (Emma Kabenge EWR=2)

It was not always possible for elders to maintain an adequate water supply. Some grandchildren were not reliable:

I don't have any water to drink! My granddaughter went to school without bringing it. (Edith Manube EWR=29)

Or a combination of problems could interrupt a household's water supply:

I had a daughter who was married to a man in this village and she used to get us water, but she has left her husband and married again in Lusango. That was three weeks ago, and since then I have been paying a neighbour's son Sh100 per jerry can to bring water from the spring, but I have to go and find him every time I want some. Now my wife and I are both ill, and we have a problem because we can't fetch water. Last night we didn't eat because we didn't have any water to cook in, and we didn't have lunch yesterday either! This morning I went out to find someone to get us water, and I met Waswa, my grandson, and he got some for us, so we are going to eat *matooke* and groundnuts – we are hungry! (Noel Katusabe EWR=9)

Food preparation was done outside the house, by women and girls. Bananas were peeled and the peels returned to the plantation as mulch, while other peelings were fed to pigs and goats if they were kept. After peeling, vegetables were not washed, but stored in the salted water they were to be cooked in until they were placed over the wood fire. Cooking was done in aluminium pots, and food usually eaten with the fingers from aluminium or enamel plates. Meals were taken sitting on the floor, often outside the house. Dirty dishes were washed in water when it was available, and scoured with earth when it was not.

Those who lived a long way away from the water sources, or who had trouble ensuring the continuity of their water supply were concerned to use it only sparingly:

> What can we do? We use what we get as slowly as we can, and often don't wash even though we have soap in the house – we usually wash on Saturdays. (Ephrance Nabakka EWR=14)

Scabrous skin rashes were frequently visible on the bodies of both the elders and members of their households. Even if water was available soap could be in short supply, or the task of washing clothes beyond the ability of the elder:

> I no longer have the strength to wash my own clothes. Sometimes my granddaughter helps me, or I just wear them dirty, like today. It's a long time now since they were washed. (Edith Manube EWR=29)

Several participants complained of intestinal worm infection, a condition which is associated with inadequate water supply and poor sanitary conditions. Research has shown that increasing the quantity rather then the quality of water supplied can improve health dramatically (Cairncross and Feachem 1993), but those who relied on others to bring water to their home had very little possibility of increasing the quantity available to them.

Water quality

Intestinal worm infections can be avoided, at least in part, by drinking only boiled water. Rutishauser (1963) reported that drinking water was never boiled in Baganda households, but some people, both rich and poor, do so today:

> We keep separate cans for drinking water and water used for cooking and washing. And we boil our drinking water, as it contains worms when it comes from the spring. (Ibrahim Sabiiti EWR=8)

Some drank unboiled water, even though they knew this was unwise:

> We don't separate our drinking water, and we definitely don't boil it! We got fed up with doing it; even if you do boil it the children still take the unboiled water if it is nearer to them. (Ephrance Nabakka EWR=14)

Others were unable to boil water as they were short of firewood, or did not accept the need to boil water:

> No! That's just for white people! We just take it unboiled, cold, just as it has been brought in the jerry can from the spring. If someone wants a drink of cold water, he takes a pot, washes it well, and pours water into it. Then, whenever he is thirsty he takes a cup of water from that pot. (Julius Kwesige EWR=20)

Sanitation

All houses were required by law to have a pit latrine, and one member of the local council was charged with the responsibility of ensuring this law was observed. In Kikole, all but three elders' homes have latrines, but many are in poor repair. The alternative site for defecation is usually the banana plantation close to the house, where the faeces can be hidden in the leaf mulch which covers the ground. This moist, permanent ground cover provides excellent conditions for hookworm development (Bradley 1975).

Houses do not contain rooms designated as bathrooms. Richer people have an outside block containing a pit toilet and a room in which to wash in privacy. Some households have a pole and leaf shelter close to the house which provides some privacy when washing. Most participants had no specific building for washing and washed inside their houses, or outside, at night, when there were few people passing by, and darkness provided privacy.

Accidental injuries

The village environment contained many hazards. Some of these were the result of the deterioration of simple buildings, which were vulnerable to the attacks of termites:

> My house fell down on me! I had to be dug out and carried away like a baby! Ever since then I have had a bad back. (Ester Namukisa EWR=27)

The participants were vulnerable to accidents in the course of their daily activities:

> We don't have the energy to look for firewood. Yesterday my wife fell down while she was looking for some – look, you can see her bruises! (Vincent Katorogo EWR=18)

> I was sleeping outside and a banana tree fell on me! Ever since then I have had a pain in my back and side and can't sleep. (Proscovia Nanyonga EWR=26)

Kjellstrom and Rosenstock (1990) note that, as the epidemiological transition progresses, pre-industrial hazards such as unsafe drinking water, poor sanitation, vector-borne diseases, and agricultural accidents are replaced by modern dangers, including urban air pollution and traffic accidents. There was very little evidence of this change in Kikole, but several traffic accidents involving bicycles occurred during the study period, one of which involved an elder:

I was knocked down by a coffee buyer on a bicycle! That was two weeks ago and my joints and back still hurt. (Christine Nabagareka EWR=23)

The diet of elders in Kikole

The trouble is bad diet. When I was young I ate well – fried meat, beer, whatever I wanted, but these days I eat badly. (Vincent Katorogo EWR=18)

Malnutrition, or undernutrition, are widely reported in developing countries. In Uganda in a recent survey of rural households only 24% reported always having enough food to meet their needs, 39% reported occasional food deficits, and 37% said they experienced a permanent shortage of food (Statistics Dept 1996a). In the same study, anthropometric surveys show that 40% of rural children under four years of age are stunted. As Eveleth and Tanner (1990) assert, the heights and weights of a nation's children reflect the average nutritional status of its citizens, and since, like children, the frailest aged are dependent on others for their food supply, it is likely that they, too, are inadequately nourished.

It was very difficult to monitor the diet of study participants, since it was not possible to observe their intake as often as would have been necessary to obtain sufficient data to make a reliable quantitative assessment of their nutritional intake. However, to gain an understanding of their dietary patterns they were asked what they had eaten the day before our visit and were intending to eat on the day of our visit. They were also asked if they had eaten meat since we last called.

Kiganda meals consist of "food" and "sauce". "Food", in this context, means "staple" and "sauce" refers to the accompanying dish. This may be meat (beef, pork, goat or exceptionally, chicken) or fish (fresh or fried), fried or boiled vegetables (eggplant, beans or peas, or groundnuts), or when these are in short supply, *ddodo*, *kakayira* or *ntula*, wild leaves and berries which grow on unused land. For the majority, who eat meat and dairy products only rarely, wild leaves are an important source of iron and calcium (Fleuret 1986). Meals are taken with a great deal of salt, if it is available.

The word for "food" in *Luganda* is also used for the most popular staple, the plantain *matooke*. According to *Kiganda* myth, *matooke* was brought to Buganda by Kintu, the first *Kabaka* (Roscoe 1911). In some versions of this myth Kintu brought *matooke* with him when he came to Buganda from heaven (Ray 1991), making it quite literally "the food of the Gods", which was prized above all other foods:

...it is believed that there is no better food for the stomach than *matooke* and, as one of the main reasons for eating is to fill the stomach, the large scale consumption of *matooke* is logical. (Rutishauser 1963 p138)

As discussed above, declining soil fertility, together with the elders' reduced physical abilities are leading to reductions in their crop yields. The yield of *matooke* is now smaller than it had been in the past, causing concern to the old people since, as

Robertson (1978) notes: "..for the Baganda a meal without *matooke* was a meal without food." (p143)

The desirability of this food can be further seen in Roscoe's report: at the turn of the century, chiefs ate three meals of *matooke* a day, but peasants only two (Roscoe 1911). Mair (1934) notes that the word "hunger" was used to describe the experience of not eating *matooke*. This comment is still applicable today:

> We eat cassava when there is famine – when we have no bananas. Our soil is not suitable for banana trees, and they go a long time without producing a bunch. Even when they do the bananas are very thin. So people with a banana plantation eat cassava when there is no *matooke* ready to eat. Those without banana plantations buy them from those who do have them when there is plenty, but otherwise they eat cassava. (Rebecca Nambiru EWR=22)

There was a general agreement among the elders that *matooke* was the most desirable food, and meat the most desirable sauce. In the past, peasants sometimes went several months without eating meat (Roscoe 1911), and this remains the case today:

> We eat meat occasionally, maybe once every three months. And even then I will only have been able to afford half a kilogramme, which isn't enough for a family this size. (Henry Ssewanyama EWR=15)

Meat was expensive, at Sh2000 a kilogramme, and smaller amounts of money were used to supplement the diet in other ways. Sugar was a very popular luxury:

> I eat meat very rarely as I don't have money, and if I do have money I buy sugar and put it in the tea I have with my cassava. I eat meat at Christmas and Easter. (Christine Nabagareka EWR=23)

> Beans are good for poor people, but rich people can eat meat and fish and sugar. We'd like to but we can't afford it. (Polly Katate EWR=7)

Dried fish was popular as sauce and could be purchased relatively cheaply. A small dish of these cost Sh300:

> People with money are eating *mukene* or *mukede* [dried fish] these days but I don't have money. (Christine Nabagareka EWR=23)

Milk was only available to the few people who kept their cows in the village, or those who were able to buy it from these people.

Food supply

A survey of agricultural methods and productivity in Masaka and Rakai districts by Hunter *et al.* (1993) found widespread declines in household food production in the period 1986-1991. There were a number of attributable causes: poverty in old age which prevented the hiring of substitute labour and the purchase of tools or other agricultural inputs, a decline in soil fertility, and a loss of labour due to sickness and death of family members, frequently from AIDS.

In Kikole, while climatic variation, the condition of the land, the increase in plant pests and diseases, and the extent of individual's knowledge of how to tackle these problems all influence production, elders have specific difficulties in attempting to maintain crop production. Primarily, these are a result of their declining physical ability or increasing frailty. Some, for example, were able to work only for short periods, while others were afraid to work in the heat of the day. As a result the area of land which they could maintain under cultivation, and therefore the size of the crop they could produce was decreasing:

> The area in front of the house next door is mine. It was cleared once, and has coffee in it, but it is bush again now as I am weak and cannot dig it. (Solomon Nsamba EWR=13)

> I should be digging now so that I have the land ready to plant seeds next season, but I don't have the strength to do that. (Polly Katate EWR=7)

The agricultural workload in Kikole varied markedly between seasons, and at times when it was elevated elders' physical limitations meant that some tasks were completed behind schedule, with consequent effects on production. These difficulties were not confined to the growing of crops, as keeping animals also became a problem as one became less strong. Today, most participants have fewer animals than when they were younger:

> I had a few cows which I kept on ropes, but when I could no longer look after them I moved them to my son, the one who died recently. Most of them died as well, and I have only one left, and it is being cared for by one of his orphans. (Grace Nanteza EWR=6)

Todaro (1989) notes the negative impact rural-to-urban migration of household members and the education of its children may have on the rural household labour supply, particularly at the times of year when workload is high. Some elders who owned a large amount of land experienced problems keeping it under cultivation, even though they were still healthy. One reason for this was a reduction in the amount of labour available, a result either of household members' economic migration to urban areas, their leaving home to start secondary education or, increasingly, their being sick or having died from AIDS.

As wages fell and prices rose dramatically after the overthrow of Idi Amin, labouring in Buganda became a less attractive proposition for Banyarwanda and Barundi migrant labourers (Jamal 1991), and an additional loss of labour occurred when the supply of these labourers ceased during the 1970s. Although this took place over 25 years ago, some larger farmers had not recovered from this loss of labour, and had been unable to maintain all their land under cultivation since then:

> I had a hut on my compound where ten porters slept. Most of them were working on my *kibanja*, so it was a big loss when they stopped coming. The government said we should rely on our own work, and do it ourselves instead of relying on porters from other countries.

But we didn't believe them, and when one year they didn't come it was a big shock. It affected our crops and our income fell. (Ibrahim Sabiiti EWR=8)

Those poorer elders whose land generated very little cash income, and who worked for cash on others' land, faced difficulties as their strength declined but their need for cash did not. After working away from home they were often unable to maintain cultivation of their own *kibanja*:

My kibanja was all cultivated once, but as I have to work in the village I can't keep it all cultivated as well – I am too weak. (Solomon Nsamba EWR=13)

My beans are all harvested, and the crop was about two tins, so there won't be any surplus. I had to work in the village for money so I couldn't grow more. (Rebecca Nambiru EWR=22)

Only the very frail did no agricultural work, and at this stage of their physical decline some of these elders received help from family members:

There is nothing I can do for myself. I can't dig any more – that's why may plantation looks so bad. But I still prepare my food and my granddaughter comes to cook it for me! (Ester Namukisa EWR=27)

The inevitable result of decreasing availability of labour for food production was a change in food production. Declining soil fertility, a result of poor husbandry which may be associated with increasing disability in old age, further reduced food supply. Since *matooke* yields fell faster than those of cassava, the latter was the most frequently eaten food among the frailer elders. Sauces made from beans, groundnuts or peas were eaten less often, and meals frequently consisted solely of starchy food crops such as *matooke* or cassava.

Figures 5.4 and 5.5 are sketch maps of the *kibanjas* of Edith Manube (EWR=29) and Grace Nanteza (EWR=6), and illustrate the declines in quality and quantity of production that occurs in old age. Edith is very old, and lives, sometimes, with a very young granddaughter, and at other times alone. Grace is younger, lives with three secondary school-aged orphans, and employs labour to cultivate her six acres. Particularly notable is that Grace grows a much wider variety of crops than Edith. Among the extra crops that Grace grows are areas of seasonal, labour intensive, nutritious crops such as groundnuts, beans, and maize, while a portion of her land is left fallow as part of her crop rotation system. In contrast, Edith's land is almost entirely occupied by permanent crops: sorghum, coffee and bananas. Their yields are slowly falling as her strength declines. She can no longer cultivate seasonal crops, and while Grace is able to support several orphans with the produce of her land, Edith is largely dependent on her nearby daughter and son-in-law for her food supply:

If you'd come here when I was younger you wouldn't have found it as it is today – I used to keep all the weeds down and it looked smart! I grew beans and ground nuts when I was still powerful, and I didn't have to worry about being hungry. But these days I am weak and have to rely on others for food. (Edith Manube EWR=29)

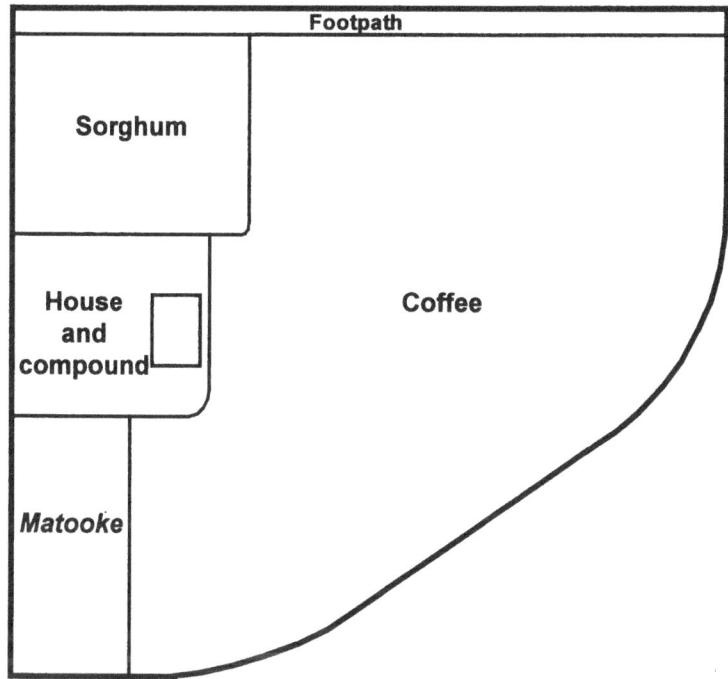

Figure 5.4 **Edith Manube's (EWR=29)** *kibanja* **(not to scale)**

Obstacles to the consumption of available food

Aside from the problems growing or purchasing food, a number of factors inhibit its consumption. Some had poor teeth, whilst others had lost their teeth, and two old women had goitres.

> My goitre is sometimes so painful that I can't eat....Yesterday I was given some cassava but it was hard and I didn't eat it. (Caroline Muguzi EWR=24)

> When you are old you get problems with your teeth and they fall out. Then you are like a young child and you can't chew your food, especially cassava. And if you can't chew you get problems in your stomach. (Gilbert Kasibante EWR=5)

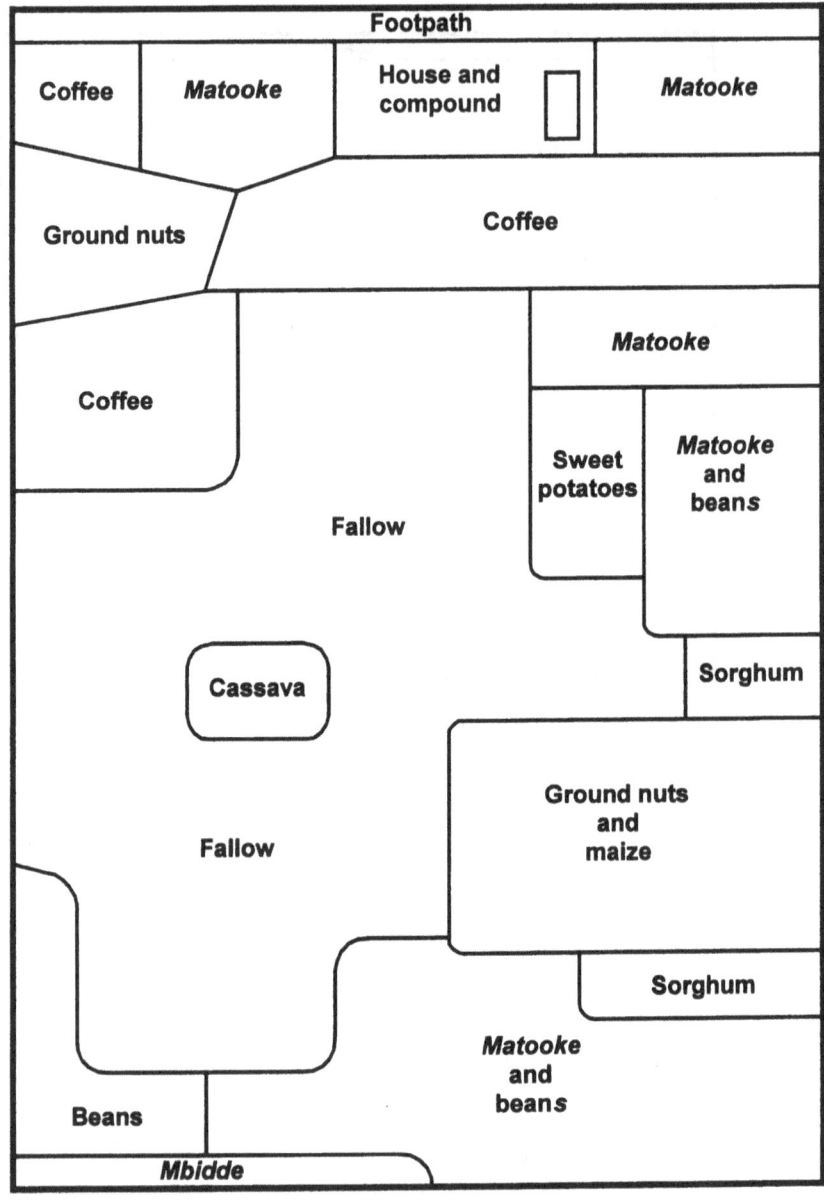

Figure 5.5   Grace Nanteza's (EWR=6) *kibanja* **(not to scale)**

There are taboos against the consumption of the animal that is the totem of one's clan (Southwold 1965). Nsimbi (1964) reports that those who did so would get a "terrible eruption of the skin". In Kikole this was said still to occur:

> If you eat your totem you can die, but more often you get a rash on your body. It is called *bigenge* [lit: leprosy]. (Young man)

Women's nutritional status was potentially compromised by a taboo against their consumption of chicken:

> It is a *kiganda* custom that women don't eat chicken. I don't know why but we have to follow our customs! It used to be eaten by men and if you came across a woman eating chicken that meant she had stolen her husband's sauce. In the past it was pork, as well, but women eat that now....Men can eat what they like! (Florence Mugasha EWR=11)

As their strength failed, and the variety of crops they could grow declined, individuals had fewer choices of food. For some, the monotony of their diet became a disincentive to eat.

> Nothing except cassava – I'm fed up with it! (Caroline Muguzi EWR=24)

Just as water shortage limited food preparation, firewood availability was also a problem. One woman, although she had food, was unable to cook it:

> Sometimes I don't eat any supper. Last night my [coresident] granddaughter went to her mother to get food because we hadn't got any. She came back with some cassava, but we couldn't cook it because we hadn't got any firewood. (Edith Manube EWR=29)

Those who were entirely dependent on others' labour could not be sure of eating:

> When my grandchildren are around they bring me cooked food, but last night they weren't here so I went to bed without eating. Since I can't feed myself any more I have to rely on them. (Ester Namukisa EWR=27)

Intra-household inequality in the distribution of available food was not mentioned during the study period, and I did not observe this at any time, in spite of being present during the consumption of many meals during the study period.

The health consequences of inadequate nutrition

Since there is a direct relationship between the quality of diet and the ability to labour (Spurr 1990), it can be hypothesised that, without assistance, the physical decline of the aged takes the form of a downward spiral. Each decline in ability results in a decline in crop production, and therefore of dietary intake, which then reduces ability further. Poor nutritional status renders one vulnerable to many infectious diseases, including malaria, tuberculosis, influenza and to intestinal parasitic infections

(Chandra 1983), all of which are commonly reported in Uganda and Kikole. It can therefore be further hypothesised that inadequate nutritional intake accelerates elders' physical decline by increasing their vulnerability to disease.

Traditional health beliefs in Kikole

The early anthropologists of East Africa paid particular attention to the nature of healing systems and of traditional medical practices, including sorcery, witchcraft and oracles (eg Hobley 1910, Evans-Pritchard 1937). Roscoe (1911) reported that, in Buganda:

> Death from natural causes rarely presented itself to the native mind as a feasible explanation for the end of life; illness was much more likely to be the result of malice finding vent in magical art. Death was the outcome of sickness which the skill and art of the medicine man had failed to overcome. (p98)

Roscoe describes a number of illnesses that are caused by the breaking of *kiganda* taboos. Some of these were associated with the rituals surrounding birth and death, and others avoidance taboos, of people and of foods (Roscoe 1911). These illnesses were reported, by the village *musomezi* [traditional healer] to be still present in Kikole:

> *Talo* and *busukko* are caused by witchcraft. A person gets a mixture of herbs, and grinds them to a powder, and puts it across the road while saying the name of a person whom he wants to harm. Then when the person steps over the powder he becomes ill. *Obukko* is caused by getting in contact with your son's wife, or a son-in-law having contact with mother-in-law. This is illegal according to our culture, and they get sick because they have broken a taboo. If they come to me I can give them a herb for bathing to make them better. (Traditional healer)

Obukko, described by a male participant in Box 5.1, is particularly relevant to this study since it limits the contact between aged parents and their adult children. It therefore influences the support given to elders by their children, an issue I will explore further in the following chapter.

Biomedical illness

Since the elders were experiencing biological ageing in the context of the poor environmental and dietary circumstances described above it is not surprising that they also experienced biomedical health problems. Poverty prevented most of the villagers using mainstream health services and their illnesses were not diagnosed and recorded. Reliable morbidity and mortality figures for the Kikole population are therefore unavailable.

Box 5.1 *Obukko*

I am not allowed to sleep in my son's home. If I did I might get an illness of shaking, the whole body, and an endless running stomach. You shake so badly that you can't pick anything up, and don't have enough blood and the skin goes pale. You get weak in the whole body, and you have to die. That is called *obukko*. If I go to there I mustn't get close to his wives. I wouldn't get it if I just stayed one night, but if I stayed a few days I would. Similarly my wife, when she was alive, would have got it if she had stayed there. You can get it from either a son or a daughter who has been married. When I go to visit my son I announce myself by shouting, or telling one of the children to warn their mother that I am coming. I just stay in the sitting room – I don't go to the other rooms. This will stop them accidentally touching me and then getting *obukko*. Then they can open the door, so that I can enter and sit. His wives will kneel some distance away from me, and greet me, and when I have had what I need I go away.

In Uganda in 1990, the main causes of outpatient morbidity were malaria (23%), upper respiratory tract diseases (16%), trauma and injuries (9%), intestinal worms (8%), and diarrhoeal diseases (7%) (Ministry of Health 1991, cited in World Bank 1993b). The pattern of illnesses reported by participants in Kikole is broadly similar. Acute infections were usually described as "flu" (using the English word) when mild, and "*musujja*" [malaria, fever] when severe, and these were the two conditions most frequently reported by outpatient attenders, as mentioned above. Kengeya-Kayondo *et al.* (1994), after a study of malaria in a nearby population came to the conclusion that although "*musujja*" literally means fever, it was most frequently used to describe an illness which appeared to be malaria. However, the participants in their study did not all understand the link between mosquitoes and the disease, and comments made by elders in Kikole reveal that they also did not understand the link between mosquitoes and malaria. For example, Rebecca Nambiru was of the opinion that boiling water was necessary because "it contains germs that give us malaria".

In 1965, Brown and Wilks (1966) found hookworm infestation in 44% of primary school children in Masaka town. In Kikole, chronic illness was frequently attributed to *njoka* [intestinal worms]:

> When you have worms you vomit or get a skin irritation after eating *matooke*. It will go away quickly, say in about two hours. It is usually an indication that the worms in your stomach don't like the food you have eaten. They can stop you eating everything, not just bananas, and you can die! (Florence Mugasha EWR=11)

Other chronic conditions frequently reported by the participants included sore eyes, failing eyesight, headaches, skin irritations, and muscle and joint pains. Chronic illness was additionally and frequently attributed to "[blood] pressure", or to "not enough blood". From a biomedical viewpoint, blood pressure was quite possibly the cause of some problems, given its high prevalence among older people in other populations, and since the local diet contained very large amounts of salt. In view of the lack of meat in the participants' diets "not enough blood" may well have been synonymous with iron deficiency anaemia. This could also explain the dizziness that was often reported to be associated with exercise and "lack of blood":

> I have a swollen face because I don't have enough blood, and I don't go to burials much any more because I get dizzy when I walk long distances. (Rebecca Nambiru EWR=22)

Some of the language used by participants is reminiscent of those who hold humoral nutritional beliefs; of the importance, for example, of maintaining a balance between "hot" and "cold" foods (Harwood 1971). These beliefs were not noted by either Roscoe (1911) or Mair (1934), although Bledsoe and Goubaud (1985) describe similar beliefs among the Mende of Sierra Leone. I did not find them expressed in any context other than in connection with one's blood, and always in relation to diet. Some meats, red herbs, fish and milk were said to be able to replenish blood, and thus one's strength, after illness:

> I have been very ill with malaria. They had to take me to hospital on a chair strapped to a bicycle! They told me at the hospital to eat millet porridge, so I have bought some of that, and passion-fruits and meat and eggs. Also, I have been drinking milk from my cows, and I have bought *mukede* and *tilapia* [fresh or smoked fish] to regain my blood. (William Muwonge EWR=3)

> The medical people told me that I should eat *mukene* as it is good for the blood, but I don't have enough money so I will use that a red herb that grows in my *kibanja*. If I boil it and drink the red water I will regain my blood. (Rebecca Nambiru EWR=22)

It is clear from the participants' statements that their current concepts of health and illness contain elements of both traditional and biomedical belief systems.

Health seeking behaviour

The health seeking process can be divided into two phases (Chrisman 1977). First there is an evaluation of the symptoms of an illness, followed by the development of an appropriate plan of treatment. Local beliefs concerning illness causation will influence the former, while the latter will be developed with reference both to the results of the former and to the context within which it is experienced. This context includes the circumstances of the sick individual or of those responsible for him or her, and the attributes of the available health services, including their cost, accessibility, and quality.

Kleinman (1980) has proposed the "explanatory model" as an aid to the understanding of the individual's conceptualisation of illness and health care choices. Individuals have explanatory models which reflect their own personality, and their social and cultural environments. While biomedical practitioners' models tend to be rigidly attached to the biomedical model of disease causation and illness, and therefore to be broadly similar, those of a spiritual healer are a product of that person's beliefs concerning one or more other possible causes of illness and therefore may vary between healers. Patients' models, in turn, will be a product of their knowledge, beliefs and circumstances, and will influence their treatment choices. Thus, any interaction between health practitioner and patient is a product of both their explanatory models.

The explanatory model may prevent a patient seeking health care at all, and Helman (1994) suggests that a decision not to obtain health care can be made not only on the explanation for the cause of illness, but also on the ability to access or pay for care, or on the expectation of economic or social problems arising from the decision to accept care. Thus, someone may underestimate the seriousness of an illness if they are unable to pay for care, or are concerned that, after paying for care, the well-being of their household or family may be compromised. This behaviour has, for example, been reported among people with cancer in Thailand (Sutnick *et al.* 1984, Bennett *et al.* 1994). The participants, most of whom had difficulty paying for health care, were frequently dismissive of their health problems:

> The disease of old bodies are chronic; we complain of back pain and so on, and even in the dispensaries they can't treat them...I've had back pain for a long time. I tried tablets and injections but they didn't help, and I spent a lot on local healers. The nurses told me that my pain was due to old age, so I decided to give up and accept death if that's what happens. (Christine Nabagareka EWR=23)

> My health's not bad; I have blood pressure and leg pain, but that's just old age. (Emma Kabenge EWR=2)

Whether they classified illness as a product of ageing or otherwise, participants had individual explanatory models, which they used to classify illness. Most people were of the opinion that some illnesses could only be cured through spiritual intervention by a *musomezi*. For them, *talo*, *busukko*, and *obukko*, all fall into this category, as does mental illness. Vincent Katorogo, however, had a low opinion of traditional practitioners, and his explanatory model, perhaps further influenced by the doctrine of the Church, excluded them completely:

> I have always disliked them. They ask you for a goat, or a cow, and even then you may not get cured, so I don't consult them. And the church tells us not to go to them, too. (Vincent Katorogo EWR=18)

In spite of the existence of these spiritual illnesses, none of the participants reported the need to consult a *musomezi* during the study. All health care choices and treatments described related to the use of biomedical health services and treatments, or to self-treatment with herbal or biomedical remedies. Although I was not informed of spiritual treatment being obtained, and do not describe any such instances here, this may not indicate that these events do not occur. Mair (1934) reported that traditional spiritual practices were rarely conducted openly, and disapproval by the Church may perpetuate this behaviour in Kikole today, where a dying person who is known to hold traditional beliefs is denied the support of the Church:

The family call the lay preacher to anoint the sick, and if he thinks there may be a traditional shrine somewhere in the house, he will first make sure the old man or woman finishes with it and throws away everything that concerns the African traditional religions. (Roman Catholic parish priest)

A "bonesetter" also practised in the Kikole area. His practice was acceptable to the Church, and the participants were ready to discuss their experience of this treatment:

> I have severe back ache and haven't been able to move much for a long time. I've had a bonesetter come and try to reduce the pain....I had to buy some cow's butter, and he mixed some herbs in it and rubbed it on my back. But I am still feeling some pain. (Proscovia Nanyonga EWR=26)

Having decided on recourse to either spiritual or biomedical treatment, one had next to decide where to go for treatment. This decision was made with reference to the relative wealth of the household and perceptions of the abilities of the practitioners among whom the choice was to be made. If one's initial assessment was incorrect and subsequent treatment therefore unsuccessful, alternative treatment would be sought. This could occur in cases of both traditional and biomedical illness, suggesting that the initial assessment is not straightforward, and therefore that it is not easy to draw a distinction between traditional and biomedical illness:

> I stepped on some herbs which had been thrown on my compound to make me ill, and later my leg swelled. I used local herbs and went to the healers but that didn't work and in the end I went to the hospital where they cut my leg and some pus came out. (Julius Kwesige EWR=20)

For those with the resources to purchase biomedical health care, individual perceptions of disease causation and severity influenced the choice of initial consultation:

> There are some qualified people in the village – Emma Kabenge, for instance [a biomedically trained midwife], but if the illness was more serious I would go to Ssematiko's clinic or to Kyamulibwa [government] dispensary. (William Muwonge EWR=3)

> If I am sick I send John [a neighbour] to buy tablets from the shop. If it's more serious I can send for one of the injectionists in the area, or I could go and consult Emma Kabenge. (Christine Nabagareka EWR=23)

The cost of health care to poor people in developing countries has received much attention in the literature (eg Corbett 1989, Abel-Smith and Rawal 1992, Over *et al.* 1992, Sauerborn *et al.* 1995). For the elders of Kikole, the choice of treatment is, like so many of their life choices, usually an economic decision: "What can I afford?" rather than "What do I need?". Since biomedical clinics do not give treatment on credit, money must be raised before treatment can be obtained. For this reason local herbs are often tried, without cost, before going to a clinic:

My wife has a cough. We tried local herbs but it is no better, so I picked this little amount of coffee which I am drying and want to sell so I can take her to the clinic. (Julius Kwesige EWR=20)

Some people chose to go to health workers with whom they had a family or social connection, and who would provide treatment on favourable terms:

I got some tablets from my daughter who runs her own clinic in Kikangazi. When we need more tablets I send a grandson to get them from her. (Michael Kavuma EWR=3)

I went to a man in Kamuli. He is the husband of my granddaughter – a daughter of one of my sons who died. The government dispensary won't give me credit and they only give you a partial dose of medicine, but that man gave me Sh6000 credit and a full dose! (Proscovia Nanyonga EWR=26)

In this final comment, Nanyonga raises an important factor in her decision making: her perception of the quality of the health care she is purchasing. Kironde (1985), notes that, in rural Buganda in the early 1980s, it was necessary to pay so much in bribes to government health workers that few people used the government dispensaries. This is still the case in Kikole today:

We used to go to Kalungu [a government dispensary], but these days there aren't enough drugs and they want a lot of money. So I went to Kabale [also a government dispensary] and they wanted money too; so I didn't get treatment....I don't know what's happening to the government clinics – the people there want to be bribed, and if you don't have money you won't get medicine. (Noah Sserwadda EWR=12)

At the government dispensaries you pay Sh400 but you don't get a full dose. Then they make you go back tomorrow, and the next day, and every time you have to pay! (Ephrance Nabakka EWR=14)

Biomedical treatment was, instead, obtained from one of the many private practitioners or facilities in the area:

If you go to a private practitioner you don't need to bribe them and they give you a number of drugs so you only have to go once. You don't have to go back and forth from home to the clinic, and when you add up the total cost of government treatment it is equivalent to the cost of going to the private practitioner, and without having to travel, so why go to the government clinic? (Noah Sserwadda EWR=12)

The person at the government dispensary also has a private clinic. If you go there you see the same person! So we go and see them privately and that way we avoid going many times. The cost works out about the same. (Ephrance Nabakka EWR=14)

Private practitioners varied widely in both the services they offered and the fees they charged. Those mentioned above were the cheapest, and one or more of these practitioners was consulted by most elders during the study period. At the other extreme, Vila Maria hospital, run by the Catholic church, and approximately 25km south of Kikole, was said to offer the best inpatient treatment, but was beyond the economic resources of all but a few participants. In between, and a popular choice for those with limited resources, was a private clinic run by Emma Kabenge (EWR=2), one of the village elders, and a qualified and highly experienced midwife, now retired from government service. She had a purpose-built clinic, containing a delivery room and a recovery room, next to her house in the village. Here she provided a relatively safe environment for women to deliver their children. She also provided care for people with malaria or acute infections, most frequently for children with pneumonia. She did not, however, treat the chronic conditions of the aged.

In a neighbouring village, Ssematiko, a qualified nurse, owned and ran the most sophisticated non-government health facility in the area, providing both inpatient and outpatient care. One man spent time here during a severe episode of malaria, and another paid for a daughter, who had antenatal problems, to deliver her baby as an inpatient. A third, who fell from a tree in March 1997, recounts his experience of this clinic in Box 5.2.

His comments, together with those of Kasibante, below, reveal another limiting factor on the effective use of available health care: their incomplete understanding of biomedical treatment. Without a full understanding of the treatment they were buying, it is unlikely that they could make optimal use of their resources, or those of the health services:

> I went to Ssematiko's and they examined me with that machine that looks like a watch. They put it on my chest and stomach and wrote things down on a form. I didn't know what was going on. (Gilbert Kasibante EWR=5)

Specialised medical treatment was available only in Kampala, and was beyond the resources of almost all participants. One old man, who had a wealthy son in Kampala, reported that he had an operation to correct his prostatic problem in a large hospital there three years ago. He frequently complained of continuing pain which he felt was a result of this operation.

In spite of these options many poor people were unable to obtain treatment due to lack of cash:

> I have a problem that produces a sound in my stomach, and has made one side of my body, from the chest to the legs, numb. If I had money I would go to the government hospital in Masaka, and see a doctor who would give me an injection and cure me. (Henry Ssewanyama EWR=15)

The problem of cash availability is not confined to the poor. One relatively wealthy couple had to choose which of them would receive health care:

> My husband needs tablets every week, and my joints don't work properly. They all hurt and at night I get a lot of pain. That's why I am here, sitting in the sun; I'm trying to get them warm. We can't afford to pay for treatment for both of us. (Johanna Birungi, wife of Gilbert Kasibante EWR=5)

Box 5.2 An accident

March 1997

I fell off a tree top and I was in a bad condition! I climbed up to get some jackfruits but I was unlucky and fell down. While I was here in pain my younger brother came to see me. He had disappeared for ten years! I was hoping he'd help me pay for the treatment, but he has left today.....

April 1997

My older brother took me to Ssematiko's on a bicycle. The doctor charged me Sh5000, and told me to eat well – fresh tomatoes and fish. My friend Godfrey in Kasakere lent me Sh10,000; I gave Sh5000 to the doctor, and will use the rest for food. I will have to give Godfrey five tins of coffee when it is ready in May.

May 1997

The doctor at Ssematiko's told me to rest until I feel OK, but I still have pain in my forehead and in my arm, and I feel dizzy, like I am going to fall down when I am walking. I can't do any work. My wife is doing the cultivation. She weeded and now she is planting beans. She is working alone! I went to Ssematiko's again, twice, and paid another Sh15,000 for treatment. I have borrowed that and another Sh5,000 from Godfrey. I was given some injections, and they told me that I need an operation as I have a blood clot in my head, and if I don't have it taken out I will go mad. All the time these days I am tired and sleeping in my bed. I can't work even though I had that treatment.

June 1997

I feel much better, but I am only picking coffee, not digging. If I bend down too long I still feel dizzy. When I go to Ssematiko's they put a machine on my forehead, which they say can tell what has gone wrong inside it. They say it is sucking out the blood that has clotted in my head, but I have never seen anything it has removed from my head. I don't know if they are doing the right thing or the wrong thing, as I still get pain in my head, and it is always a lot worse after I have had the treatment. I have been hoping my younger brother would come back and be able to pay for my treatment. He gave me Sh1000 to buy some sugar when he was here, but he hasn't come back yet. I haven't been able to do any digging since I fell off the tree.....

Inpatient care was beyond the reach of almost all elders. Not only were there fees to pay, but inpatients needed someone to care for them while they were in hospital, as the staff did not wash or feed patients, nor did the hospital provide food for them. Food, cooking utensils and firewood had to be taken to the hospital, with associated transportation costs. Hospital equipment was limited; it was always necessary to take one's own sheets, and sometimes a mattress. Therefore, most people opted for home treatment:

> We have been going to Kagimu for a long time. We choose him because there is no-one who could look after me if I went to hospital. My husband, Salongo couldn't do that. (Gertrude Najunjju EWR=17)

Another incidental cost of illness is that of the loss of labour to the household while a member is away in hospital, or away caring for a sick relative. This problem has become more significant during recent decades, as the HIV/AIDS epidemic has selectively killed young, productive adults (Barnett and Blaikie 1992, Pitayanon *et al.* 1994). Their absence results in a loss of production, and therefore negative consequences for household well-being. One wealthier woman was able to use her available cash to avert this problem:

> I had to use labourers on my *kibanja* while I was sick in hospital, and for a while after I got back, until I was well again. (Emma Kabenge EWR=2)

Since most health services and practitioners were situated away from the village a journey to get treatment was usually necessary. Lack of transport, or physical weakness prevented some old people obtaining treatment:

> We went to my son to tell him we were sick, but he said there was no vehicle around to take us to hospital. And since then he hasn't even been to see how we are! (Polly Katate EWR=7)

> I am still living, but only due to God's power. My ears are becoming very deaf and my whole body is very itchy. I haven't the strength to go to the hospital and get medicine. (Ester Namukisa EWR=27, who died six weeks after making this statement)

> I can't visit my friends any more because I have leg and stomach pains, and I can't find anyone to take me to Ssematiko's place to see the doctor. (George Kabongo EWR=16)

Another chose to walk to obtain treatment:

> I sent messages to my grandson in Kabuswaga three times, asking him to take me to Kamuli for treatment, as he has a bicycle, but he never even replied! Since I was in pain and there was no-one to help me I decided to walk there [at least 10km] early in the morning, before the sun was strong. Since I had the treatment the itching all over my body has stopped, but now I have leg pain, because I had to walk all the way home! I was told to go back for more medicine but I am afraid of the journey. (Proscovia Nanyonga EWR=26)

The participant who was disabled by a left-sided weakness was dependent on a neighbour to take him for his regular treatment:

> I see the doctor at Ssematiko's clinic. My neighbour pushes me there on his bicycle. I saw him last week, on October 12th. (Gilbert Kasibante EWR=5)

For those who found it difficult to buy health care, or travel to a clinic, self treatment was common. Sick people, with a little money or access to credit, could purchase tablets from one of the village shops, which carried a small stock of analgesics, antihelminthics, antimalarials, and antibiotics. The quantity purchased was limited by their resources, rather than the amount needed to produce a complete cure. They were dependent on the advice of the shopkeeper and the availability of drugs at his store when deciding what to buy. In 1987 there were only 45 pharmacists in the whole of Uganda (Whyte 1991), and the shopkeeper was therefore unlikely to have more pharmaceutical knowledge than his customer:

> I feel that I have blood pressure and my heart feels unsettled....I bought some tablets and now I feel better. I told the people in the shop what my problem is and they gave me the tablets they thought I should have. The amount they give you depends on the amount of money you have, and if you have a lot of money you get a lot of tablets. I didn't take them all at once though, I took two a day until they were finished. (Juliet Kisaakye EWR=25)

Although their knowledge of biomedical treatment was limited, aged people were said to have a good knowledge of traditional herbal treatments, and their home use was frequently reported. Herbs to treat both spiritual and biomedical illness can be self-prescribed and purchased "over the counter":

> I take *mmumbwa* [a herbal remedy] for my strong heart beat – here it is! It's made of clay with herbs in it and I grind it into a paste in this special bowl, and then drink it. I bought it in the market, and you can get others there for different problems, including *talo* and stomach pains. (Florence Mugasha EWR=11)

There is a body of traditional knowledge concerning the use of herbs in response to biomedical disease. Bitawha *et al.* (1997), after studying herbal treatment of malaria in Uganda pronounced it effective, and in Kikole herbal treatment for malaria was said to be efficacious and was widely used:

> For fever, I collect a herb called *mululuuza* and cook it and give it to the patient. It is effective. (Michael Kavuma EWR=3)

Traditional herbal treatment also provided temporary relief for stomach pains:

> Muwonge is sick some times and complains of stomach pains. I think he has worms. I was told to use *kisanja*, a herb which grows in my compound. I cook it and give it to the child, and he gets diarrhoea, but then is better for a while, but the pain always comes back. (Rebecca Nambiru EWR=22)

Health information

During the study period no health workers visited Kikole to provide health education
or information, but those participants who went to clinics would be given advice on
the management of their condition. Since most villagers are illiterate, other health
information was most commonly obtained via the radios of the most wealthy, and
then disseminated by word of mouth. There was no health information directed
specifically at the needs of the aged. Seminars on the prevention of transmission of
HIV had been held in the past, and old people's attitudes to these are discussed in
Chapter 7.

Summary

Most old people in Kikole experience health problems. Some of these are the result of
poor nutrition, others of living in an unhealthy environment, and still others may be a
part of their biological ageing, although this should not be assumed to be the case.
Individual health care choices are influenced by one's beliefs on disease causation, the
severity of symptoms, one's ability to purchase treatment, one's ability to travel to a
health facility to obtain that treatment, and by local perceptions of the quality of
service offered by the health care providers. Although traditional spiritual and
physical, and modern biomedical theories of disease causation are all held to be valid
in Kikole, and health practitioners are available to provide treatment for illness with
either a spiritual or biomedical origin, none of the participants reported choosing,
during the study period, to consult a traditional, spiritual healer.

Those whose health or physical ability deteriorates as they grow older may have
difficulty maintaining their well-being without the support of others. In Kikole, the
only sources of such support are one's family and fellow community members. The
extent of the provision of support is influenced by the quality of the relationship
between provider and recipient. In the following chapter I explore relationships
founded in both kinship and friendship, and discuss their significance in the provision
of support to the aged in Kikole.

Chapter 6

Elders' Social Status and Relationships

Aging and modernisation: "Dependent nations; dependent policies; dependent old people".
(Neysmith and Edwardh 1984 p34)

This chapter is concerned with the ability of older people in Kikole to maintain their well-being in the context of the social, cultural and economic changes that accompany "modernisation", as defined by Cowgill:

> Modernisation is the transformation of a total society from a relatively rural way of life based on animate power, limited technology, relatively undifferentiated institutions, and parochial and traditional outlook and values, towards a predominantly urban existence based on inanimate sources of power, highly developed scientific technology, highly differentiated institutions matched by segmented individual roles, and a cosmopolitan outlook which emphasises efficiency and progress. (Cowgill 1986 pp186-7)

If the indicators contained in this definition are applied to Kikole, a peasant community which is largely self-sufficient in food, where there is not a single motor vehicle, almost universal severe poverty, and very limited contact with the wider world, then the village will be considered to be relatively unmodernised. However, this would be an incomplete picture. A large number of the village's young people, demonstrating behaviour consistent with advancing modernisation, have migrated to urban areas to find work, and are very much embroiled in the modern world. The changing attitudes which have led them to leave their home village are disguised by these migrants' absence. Similarly, the relatively undeveloped peasant economy of Kikole is an exploited component of the national capitalist economy (Bernstein 1979, Harriss 1987), and has been so since the village was first settled. It is therefore likely to be much changed by the modernisation of Uganda. In the estimation of the impact of modernisation on Kikole, therefore, the lives of old people cannot be seen in isolation from those of their children, who are their major source of support in old age. Nor can the economy of the village, within which the aged subsist, be seen as separate from the national economy.

Neysmith and Edwardh (1984), in recognition of this relationship between community and its economic environment, have applied dependency theory (Dos Santos 1970, Cardoso 1972, Palma 1978) to the lives of the aged in developing countries. Dependency theory attributes continuing poverty in developing countries to the removal of their economic surpluses by more powerful developed countries. In Uganda the structural adjustment programme of the World Bank and the International Monetary Fund has, since 1981, been the most recent manifestation of this process (Jamal 1991), and through its insistence on tight controls on government spending, has further compromised the position of vulnerable groups within the national population, including the aged (Obbo 1991, Brett 1992). Government policies do not provide older citizens with free health care or old age pensions, while the local elites mentioned earlier (Mamdani 1976), have relatively high wages and advantageous conditions of employment which allow them to accumulate wealth and ensure their own old age is not spent in poverty. They can be described as agents of the developed economies; as willing facilitators of the removal of the surpluses created by their relatively poor rural compatriots. These village people are thus rendered unable to accumulate capital during their productive years and therefore experience insecurity during their old age. Poorer elders, therefore, continue to turn to traditional sources of support: their children, grandchildren or fellow community members. Unfortunately, they find that these people, too, are suffering due to the loss of the village's surplus production, and are in no position to offer the support their older relatives or neighbours need. The process of modernisation, which one feels intuitively should be of benefit to the needier members of society, appears to disadvantage them and to be of benefit only to those who are a part of its implementation.

Social status and modernisation

Pioneering cross-cultural research on ageing and modernisation by Cowgill and Holmes (1972), when analysed together with other work concerning the status of the aged in modernising societies (eg Simmons 1945, Maxwell and Silverman 1970, Palmore and Manton 1974, Silverman and Maxwell 1987) has led to an awareness that the primary cause of the decline in the status of the aged is their loss of control, as modernisation proceeds, over tangible and intangible resources. In particular, they have lost their role as disseminators of information which is believed, by younger people, to be useful, either socially or economically, to members of their society (Albert and Cattell 1994).

The control of the transfer of information in a traditional society is related to its political and social structure. For example, a number of East African tribes are gerontocracies, and comprise a system of age-sets through which men pass, as they move from young adult to elder status (Radcliffe-Brown 1940). The members of these institutions have specific duties within their society, with younger age-sets, for example, being the tribe's warriors, and the elders being expected to act as its advisers or politicians, and to play major parts in its ritual life (Mair 1965). Spencer (1965), in Kenya, and Rosenmayr (1988), in Mali, both note that in age-set societies the accumulation of information is formally linked with advancing age, with knowledge passed from old to young only at the predetermined moments of transition from one age-set to the next. Thus, only by becoming old can one accumulate the important cultural knowledge which will entitle one to a respected position in society. Nydegger (1983), however, wisely urges caution when assuming that societies that are controlled by an elite group of old men are necessarily societies where all older people have high status. Sangree (1987) has noted, among the Tiriki of Western Kenya, the relatively low status of childless men when compared to other men, and there are many examples of African societies where aged women, and particularly widows, have generally lower status than aged men (Potash 1986, Folta and Deck 1987, Iliffe 1987, Tout 1989, Owen 1996).

The social status of the aged in Buganda

There is no formal age-set system among the Baganda, and older men are therefore unable to protect their status through the operation of such a system. In societies such as this, the transfer of information is the responsibility of parents and other members of the parents' generation (Radcliffe-Brown 1950). In Buganda, parents and paternal relatives have, traditionally, been responsible for the education of children (Mair 1934).

In the past, high status was obtained through one's position as a chief, as an ally of a respected chief (Roscoe 1911), or through the attainment of old age (The Kabaka 1971 [first publ.1935]). Respect for the aged was demonstrated in the readiness of younger people to accept parental authority (Roscoe 1911, Mair 1934, Richards 1964). The residents of Kikole agreed that there persists today a cultural norm requiring respect to be shown to old people and parents:

The highest respect must be given to your parents. (26 year old man)

Respect comes from children and grandchildren. Those without children have the least respect. (57 year old man)

That respect may indeed be given to the aged today is further revealed in the contemporary use of the Luganda word for "old person": *mukadde*. This is used as an honorific when addressing an aged individual, and also when addressing a younger individual who is seen as worthy of respect for reasons other than age, implying that to publicly exaggerate such an individual's age is a respectful gesture.

Radcliffe-Brown (1950 p27) describes the knowledge that is of value to a traditional culture as that which facilitates the "continuity of the social order.... of tradition, of knowledge and skill, of manners and morals, religion and taste". This description accords closely with the content of the traditional education given to Baganda children (Mair 1934). Arth (1968), commenting on Shelton's earlier work with the Ibo in Nigeria (Shelton 1965), described, most evocatively, the roles of the Ibo elders as "...the arbiters of dispute, the libraries of tradition, and the storehouse of proverbs" (Arth 1968 p242).

These roles identify the older members of that tribe as having both knowledge and wisdom, attributes which were also said, during group discussions, to be possessed by old people in Baganda:

Being with an old person is like being with a book, because they saw things happen and they can tell you about them. (52 year old woman)

> It is because an old person told me the history of Buganda that I know about the past Kabakas....We know about our lineage, who is related to us, and who belongs to our clan. (37 year old man)

This knowledge, together with their life experience gave them the wisdom to arbitrate in disputes:

> If a couple is having problems they will look around for an old person they can take into their confidence. They may invite this person to come to their house and advise them, or they may visit him at home. (33 year old man)

According to the etic theory discussed above, if young people's estimation of the value of elders' knowledge falls, then the latter will experience a decline in their status (Albert and Cattell 1994). This may occur as a result of changes in the way young people are educated, and in the content of the education provided. In Kikole, children no longer receive knowledge which is appropriate for the maintenance of the social and cultural structure of their community, in their community and from a member of their kinship group. Instead they now go to a school which may not be within their community, and are taught subjects intended to facilitate participation in the wider culture and economy of a rapidly developing nation by teachers who may not be members of the same cultural group, or part of the same community as their pupils (Cattell 1989, Sangree 1997). Thus, old men and women find that their lifetime's accumulation of cultural information and farming expertise, which is itself the accumulated knowledge of many generations of their ancestors, is now of little interest to their children (Dorjahn 1989, Apt 1992). Nahemow and Adams (1974) identified this change in Buganda, where, after conducting a study among over 1500 Baganda secondary school students, concluded that "...modern education weakens the position of the aged as advisers and as repositories of the oral traditions and wisdom of the past". Using the same 1971 data, Nahemow (1987) reports that Baganda secondary school children sought advice from their grandparents, but that the proportion of these children who acted on this advice declined with increasing years of education. This behaviour indicates a continuing respect for the aged, but suggests a gradual devaluation of their role as advisers during the school-based education process. In Kikole, a similar decline in the status of aged people is evident:

> Children don't respect their parents any more. This is due to education, because when someone is highly educated they don't respect their parents. Young people no longer follow our culture, and if a parent wants his child to behave according to *kiganda* culture the child will reply that he wants white culture! (32 year old man)

> Young people don't come to old people for advice as much these days. Everyone is trying to get some money, so instead of consulting an old person when someone's animals damage your crops, you go to the village Chairman and he sends the business directly to the courts where people hope to get money [as damages] instead of sorting it out between themselves. (67 year old man)

The impact of modernisation on the status of elders may not be solely negative. Sangree (1997) reports that, among the Tiriki of Western Kenya, although the old men have lost almost all their traditional religious roles during the past eight years, they have become worthy of respect in two new ways. Firstly, their traditional role as the regulators of the transfer of land to the next generation has become more prestigious as farmland within their tribal lands has become scarce and increased in value. Secondly, among the Tiriki, elders are increasingly finding themselves as caretakers of land and children, belonging to sons who are working away from the village. Sangree interprets this as a status enhancing role, although he does not note the elders' attitude to this change in their workload. In Kikole, as I shall later demonstrate, the assumption of responsibility for one's grandchildren was more often regarded as a burden than a benefit, and while many young people were leaving the village due to the shortage of land, the shortage was so severe that the small amount a son received from his father was of little use to him. To survive he had to abandon the land, since it was impossible to conduct a life in the village and in town at the same time. The effect of the shortage of land was, therefore, the removal of an important source of support for their aged parents, rather than an increase in their status.

The supportive relationships of the aged

Sangree's (1997) examples are useful illustrations of the links and contrasts between status and well-being, since in each case a change which is reported to have increased the status of the Tiriki aged has compromised the well-being of elders in Kikole. They also beg the following question: Whilst it may be justifiable to assume that where elders are held in high esteem they will be afforded the respect that their culture requires, can one further assume that one who is respected will also be provided with support when in difficulty? There are two factors to be considered here; the readiness of younger individuals to provide aged people with support, whether in the observance of a norm, or through a more personal motivation, and the ability of the younger person to provide such assistance.

Thus, the provision of support is a function of cultural norms and intergenerational relationships, together with the economic situation of those wishing to offer this support. Apt (1988) raises the important question of the well-being of those expected to provide support to the aged, relative to that of the aged themselves. If, for example, the elder is perceived to be relatively better off than the adult child then the flow of support may continue to be from parent to child for some time after the child has reached adulthood. An ·analysis of the flow of support between individuals can therefore only be a part of a wider exploration of their circumstances and the nature of their relationships.

In Kikole, both old and young described the giving and receiving of support between generations as a two-way process, with the direction of flow of support changing throughout the lifecourse. Ideally, parents should care for their children when they are young, and children care for their parents as they grow old and frail:

My sons supported me at Easter; each gave me a kilo of meat and a kilo of rice.....We cared for them when they were young and now it's their role to care for me. (George Kabongo EWR=16)

It's like planting two seeds and harvesting six. Our parents produce children as preparation for their old age, expecting that they will benefit from our sweat and our hard work. (29 year old man)

It is apparent from these quotations that there is no expectation among either generation that one should repay like with like, that there is an economic measure applied to non-monetary support, or that support should be repaid within a specific time-frame. Thus a child's upbringing can obligate him to visit his or her parents with gifts, but no attempt is made to quantify the costs of either component in this exchange. Rather, there is a mutual acceptance that the behaviour described is appropriate, and conforms to the expectations of both parties.

In a seminal work, Mauss explores the nature of three obligations: to give, to repay, and to receive, which he describes as:

....a series of rights and duties about consuming and repaying existing side by side with rights and duties about giving and receiving. The pattern of symmetrical and reciprocal rights is not difficult to understand if we realise that it is first and foremost a pattern of spiritual bonds between things which are to some extent parts of persons, and persons and groups that behave in some way as if they were things. Food, women, children, possessions, charms, land, labour, services, religious offices, rank – everything is stuff to be given away and repaid. In perpetual interchange of what we may call spiritual matter, comprising men and things, these elements pass and repass between clans and individuals, ranks, sexes and generations. (Mauss 1970 pp11-2)

Here, Mauss reveals the complex, multi-dimensional nature of exchange relationships, within which exchanges are made with reference to past and future exchanges. The successful negotiation of an exchange will, therefore, require a mutual awareness of its historical and social context.

Moore (1978) introduces the "life-course arena" to describe the temporal context within which exchange relationships were maintained among the Chagga of Tanzania. For members of this society, where most material items such as housing, clothing and cooking utensils were short-lived, one's most reliable and long-lived assets were one's relationships with kin or friends, since these relationships included the possibility of supportive exchanges and were therefore essential resources during times of crisis. Support given would be returned at an unspecified time in the future, but not necessarily as "like for like". For example, child care provided one day may be exchanged for transport to hospital, or a gift of seed at the start of the growing season repaid with food at its end. These are not primarily financial exchanges, but a part of a friendship or kinship relationship. Trust in those with whom one enters into these relationships is therefore necessary for the system to work effectively. Moore (1978) asserts that families are participants in many of these relationships at any one time and, since they represent security in difficult times, they are actively accumulated over the lifecourse. Elders, therefore, are likely to have a portfolio of many such relationships from which they can request help as they become less able to support themselves. Moore (1978) is also of the opinion that social exchanges such as occur during time spent in conversation or beer drinking are rituals intended to maintain these relationships during periods when neither of the parties is in need of support, and thus ensure that they are available to provide support when difficulties arise.

In Kikole, most exchanges are of this informal nature, taking place between friends or kin. There are also more formal exchanges: firstly those involving the large sums of money needed to pay a brideprice or taxes, health care or education, the amount of which is more than could be obtained through an informal arrangement, and secondly, the exchange of contributions towards a community member's costs, incurred in the provision of a burial or an *orumbe*.

Informal exchange relationships over the lifecourse

Modernisation is changing the nature of supportive relationships in Kikole. In the following analysis I discuss the "life-term arena" of exchange relationships in Kikole, paying particular attention to the impact of recent social and economic change on the ability of aged people to obtain support from their family or community members.

The flow of support from parent to dependent child

In traditional society, Baganda children frequently did not live with their parents, but with a member of their father's clan (Roscoe 1911, Mair 1934). A daughter, on reaching puberty, was not able to sleep in her parents' home again, although, in 1934, Mair (1934) reported that observation of this taboo appeared to be lapsing. Later, when a daughter married, she transferred her allegiance to her husband's clan, who took responsibility for her well-being until her death. A son was helped by his father, or members of his father's clan, to obtain land and build a house, and then to buy a wife and start a family. An adult woman in Kikole confirmed that this norm remains current:

> Parents can help an adult boy by helping him buy land and by finding him a wife and paying brideprice for him. They make him independent. When a girl is lucky and finds a man to take her, her father will give her a big wedding party and things to take to her husband, and she will always remember this. (40 year of daughter of Noah Sserwadda EWR=12)

Today, while a child is young and dependent, there are many more costs to be met than there were when Roscoe was observing *kiganda* family life early this century. School fees, health care, and clothing are major causes of financial hardship for Kikole households: school fees, being required continually for each child, over many years, cause the most problems. Although the impact of providing this education on their own lives had been negative in almost all cases, the aged study participants did not begrudge their grandchildren an education:

Children must learn discipline in school rather than just wander round the village. (Henry Ssewanyama EWR=15)

They must be able to read. Then they can read the signs when they want to go somewhere. (Sulaiman Kagire EWR=10)

The costs could be considerable:

I have four sons and two daughters at teacher training college in Kampala, a daughter there at the YMCA handicraft training centre. Another is in secondary school at Kampung. Then there are another ten children and grandchildren living here and going to primary school! I pay their fees and everything else they need because it is important – an educated child will have fewer problems in life. (Amos Nduwala EWR=1)

We have to provide books, uniforms, fees, pencils. And recently they asked us to provide geometry sets in which to keep their ballpoints! So far we've only paid for one and we are saving for uniforms for the rest. (Sulaiman Kagire EWR=10)

Support of one's children does not always cease when they leave the household and become "independent". Adult sons and daughters may both call on their parents for help:

If you are a married daughter and you are sick your husband may not look after you well, so you come back to your parents' compound so that they can help you. (24 year old woman, daughter-in-law of Noah Sserwadda EWR=12)

Sometimes I go away to get a job, but I don't always find one, and can come back with nothing in my pocket. Then, if I get a problem, say a child being sick, I ask my father to help me. I say to him: "My child is sick, and I have no money – would you leave him to die?". He says: "No, I can't.", and my mother reaches inside her *gomesi* to get some money for me. Then I can take the child for treatment. (24 year old man, son of Noel Katusabe EWR=9)

As the second of these quotations suggests, a parent feels obligated to help an adult child. Adult children agreed, saying that since their parents brought them into the world it was their duty to support them. These compulsory exchanges are made without any assurance of repayment:

Lukyamuzi [an adult son with his own land and family] was imprisoned for not paying his taxes and my money was used to rescue him! I had Sh15,000 in my pocket, left over from my coffee, and now it's all gone. I don't know if he'll repay me. That's up to him. (Julius Kwesige EWR=20)

Not all fathers are able or prepared to do this, however. If resources are unavailable, or if the child is considered untrustworthy then support may not be provided:

We young men have been proved to be dishonest in the way we handle the things we get from our parents. Say an old person lends you money to start a business and the business collapses – you will just run away without telling anyone what has happened! If your father has been saving for you all his life, and you lose that money then he may have to sell the land he has been keeping for you just to repay that money he borrowed for you. So although they are interested in our future we have become dishonest and they don't help us so much any more. (20 year old man, grandson of Proscovia Nanyonga EWR=26)

A large landowner felt that young men's participation in the cash economy through trading in agricultural produce, even though they mostly had little or no land of their own from which to make a living, was a form of exploitation. This could, therefore, be interpreted as another example of resources being transferred from a parent to an adult child:

> They want to keep changing businesses; say first buying *ndisi*, then cassava, always changing the commodity hoping for a bigger profit. And at the same time they are expecting us old people to grow these for them, when we have already retired from that kind of work. They want us to dig so they can make money, and they are exploiting us! (Michael Kavuma EWR=3)

As adult children's independence increases, their parents are able to reduce the support they provide them, and their relations with other, non-related adults become more important in ensuring their security through the provision of support.

Supportive relationships between unrelated adults

Blood brotherhood   The adults with whom one engaged in mutually supportive relationships, but to whom one was not related by kinship ties, are described in Luganda as either *mukago* [blood brother], or *mukwano* [friend]. In the past, a blood brother relationship was created, between men, by the enactment of a ritual involving the exchange of blood, the details of which varied between tribes. Blood "brother" relationships were not entered into by women. The relationship was entered into voluntarily, following declarations of mutual affection; the relationship was very strong, "like a brother born in the same family" (Ibrahim Sabiiti EWR=8), and abusing it rendered one liable to punishment, which might include illness or misfortune in one's family. Relationships with blood brothers were therefore highly valued, and, through their voluntary nature, possibly more so than those with one's biological or clan brothers, into which one was involuntarily bound by kinship obligations of reciprocity. The opinions of the younger old men of Kikole today, indicate a changing attitude towards blood brotherhood:

> I don't have blood brothers – that kind of relationship is bad because if there is a problem between you then bad things can happen. Your cows can die or your people get sick. (Amos Ndawula EWR=1)

> There's no point in having blood brothers. When I am dead, instead of helping my children they would just try to embezzle the properties I had left them. They would just add them to

their own riches so that their own children would be richer when they died. And suppose you have an accident — when your blood brother comes to see you it won't be to help you, but to see what he can take from you! And he will want you to die so that he can inherit your wife. (Henry Ssewanyama EWR=15)

The incidence of blood brotherhood appears to be declining. The church disapproves of it, and as the quotation above indicates, there is an increasing unwillingness to trust others. I will return to the latter point during my discussion of formal exchange relationships later in this chapter.

Friendship

Our parents used to have blood brothers, and that was better than what happens now, which is friendship based on money. If you are friendly with someone because you both have money then he will abandon you when you are short of money. We have a saying: *"Okolerakai omukwano nomwavu kiki kyanaku yambamu?"* ["Why be friends with a poor person - how can he help you?"]. (Ibrahim Sabiiti EWR=8)

It seems, then, that supportive relationships between friends are now formed instead of, rather than in addition to, blood brother relationships, and do not involve the life-long commitment of blood brothers. Several conditions have to be met to ensure the continuance of this friendship. Firstly, each partner must be able to respond to a request for support from the other, and secondly, each must have confidence in his friend's ability to supply such support should he be so requested. Thirdly, each must also be confident that his friend, even though he is willing, is also able to repay such support, although not necessarily in kind or within a specific time frame:

All our neighbours are our friends. They are important, because if you don't have friends you can't get help when you need it. But if they don't help me I won't help them. (Polly Katate EWR=7)

Here, "all" neighbours are specifically identified as friends, indicating the persistence of the obligation to maintain good relations with one's neighbours (Nsimbi 1956, The Kabaka 1971 [first publ. 1935]). Neighbours are seen as important contributors to one's security in times of danger, such as a house fire or an attack by bandits, both of which occurred during my study period.

However, the monetary basis of friendship does not preclude the enjoyment of social interaction between friends, although even here reciprocity remains in evidence:

Sometimes one of us makes some alcohol and we go and drink at his home, or we might go drinking in Kilaru. We take it in turns to buy the booze, and Sh500 buys enough for us all to enjoy an afternoon together. (Henry Ssewanyama EWR=15)

Some women also go to Kilaru to socialise, but their domestic duties are more demanding than those of men, and so they most often meet their friends in their homes:

> My friend sometimes comes here with her mat, and we sit and weave our mats and talk until it is time to cook. I sometimes go and visit her, too, if I am not working in my garden. (Rebecca Nambiru EWR=22)

These occasions are opportunities to relax together, to exchange news and gossip, to learn of current market prices, to discuss the progress of crops, or to try to resolve family problems, all of which activities serve to cement the friendship through the exchange of non-monetary goods such as advice, knowledge, or good humour. For such exchanges to take place it is necessary that the participants have a mutual understanding of each other's lives and problems. Since this is obtained largely through shared experience, friendships form mostly between individuals of similar economic status:

> Do you think a poor person associates with rich people? I associate with people who are poor like me. They come here or I visit them. We are in the same group. (Juliet Kisaakye EWR=25)

Members of each of these groups experience similar difficulties in their lives and their expectations of a mutually supportive relationship are therefore likely to be comparable. The support they are able to offer each other will be qualitatively and quantitatively appropriate to the problem at hand. Friends are thus able to help each other in many ways, including meeting the costs of health care or school fees, of entertaining guests, or of providing extra food for a new mother or an invalid; by providing small gifts of seed or food, the loan of a hoe or of a young animal which will then produce young which their friend can keep, rear and sell at a profit; or by buying or selling goods in the market on their behalf.

When an individual experiences a decline in his or her circumstances, as may occur when old age reduces his or her productivity, the offering or repayment of a loan or favour by this person becomes problematic. Reflecting the spirit expressed in the local saying highlighted above, he or she becomes less attractive as a friend. In Kikole, this was reported to have occurred to men who experienced a loss in status, or in physical ability:

> I had many friends when I was the chief here, but since I resigned few come and see me. (George Kabongo EWR=16)

> When you lose the ability to walk around that is the end of your good life. Some people are friends because you have money or can give good advice. But when you lose your power those people say "That one has no money or future and is not useful any more". I only had a very few friends who still came to see me after my stroke. (Gilbert Kasibante EWR=5)

As the latter comment indicates, not all friends had a purely mercenary attitude to their relationship. Women, in particular, said that friendships were maintained after they became dependent, although when the expectation of repayment was removed in this way, material exchanges within such a friendship ceased:

> When someone sees that you are no longer powerful he doesn't see you as being as important as you used to be. So he sees no reason to bring you anything when he comes to visit you. When we were powerful a friend would come with a jerrican of beer and porridge made from millet, which we like in Rwanda, but these days they come with nothing in their hands and so when you go and visit them you take nothing as well. (Irene Mutondo EWR=30)

As one ages and one's abilities decline, it becomes increasingly difficult to maintain supportive friendships. Old people, therefore, turn to their adult children for help at this time.

The flow of support from adult child to dependent parent

Olukula luyonka abaana! Old people suck from their children! (21 year old man)

The normative obligation to care for one's ageing parents was frequently cited by both the participants and their children and grandchildren, with elders often adding the complaint that today's children don't support their parents as they did in the past. Nydegger (1983 pp26-7) has discussed the "Golden Age"; a period in the past when "a nuclear family with intergenerational extension provide[d] strength, love and sustenance to all its members". Many writers, after studying historical demographic data concerning pre-industrial cultures, doubt that this age ever existed in those countries now considered "the developed world" (eg Laslett 1965, Gubrium 1973, Hendricks and Hendricks 1977, Foner 1984). Others, many in Africa, are of the opinion that the "Golden Age" did exist until recently, and that modernisation has compromised and continues to compromise its existence (eg Ndeti 1987, Okojie 1988 p12, Dorjahn 1989, Apt 1996).

Phillips (1990) has studied government services for the aged in Asia and Southeast Asia. He asserts that government claims that the extended family is alive and well, and caring effectively for the aged members of their populations, are, on occasions, intended to serve as a justification for their failure to provide much needed supportive services for the aged. There is thus the possibility that the scale of extended family support is exaggerated in the interests of political expediency.

Such a motivation may be the explanation for an apparent contradiction in the policy of the colonial Ugandan government. In 1946, shortly after the 1941 Vagrancy Law had been introduced in Buganda to control the large numbers of unemployed and destitute people who were congregating in towns, the Government Department of Public Relations and Social Welfare was able to say:

> Apart from the generally low economic standard of the population, destitution in Uganda is not a major problem. The family structure still tends to act for the relief of its own members. (Iliffe 1987 p204)

A 1980s sociology text implies that nothing has changed in African families:

> Care for and respect for the elderly, especially one's parents are still very important in African families. Whatever the position of the child, he is expected to treat the elderly with great respect. Assistance must be given to make their less productive years trouble-free. (Kayongo-Male and Onyango 1984 p8)

In Kikole, as noted above, parents reported the existence of a social norm requiring adult children to support them in their old age. However, elders were not confident that their children would obey this norm, and blamed modernisation for their reluctance to do so:

> It's just chance. If a child has a job he may try and support his parents or grandparents. In the past we saw our parents as our God, our creators, but these days they don't support us. They just want modern things. (Gilbert Kasibante EWR=5)

Analysis of the supportive exchanges between parents and child is best achieved in the context of the child's lifecourse and gender. Two stages of this lifecourse are of particular significance; their early adult years before their marriage or establishment of their own household and family, during which time they have relatively few obligations to others, and the period after they start their own family, when they may have obligations to their partners and their children, as well as to their ageing parents. Since these obligations vary between the sexes, it is helpful to discuss the relevant aspects of daughters' lifecourses separately from those of sons.

The provision of support by younger, unmarried, adult children

Traditionally, daughters married at a younger age than boys, and were rarely found in their parents' household beyond their mid-teenage years (Roscoe 1911). The major return to parents from a daughter's upbringing was her brideprice, paid before her marriage by her prospective husband's family. In Kikole today, it remains broadly true that young women are rarely found in their parents' households, but their absence is increasingly due to their leaving home to look for work in urban areas, rather than upon the occasion of their marriage. Daughters who were still living at home helped their mother with domestic duties, and when the household agricultural workload was heavy they worked in the fields, either for their mother or their father. They were not encouraged to move around the village, where they would come into contact with young men, which their parents felt would very likely lead to pregnancy. Parents therefore arranged marriages for their daughters as soon as possible, to obtain their brideprice. This would not be payable if the daughter became pregnant, or if she migrated to town and established an informal marriage there. Unmarried daughters who lived at home were seen, by their ageing parents, more as a source of worry than a source of assistance. Many unmarried young women had left the village to avoid an early arranged marriage, hoping to establish their economic independence and avoid future dependency on a husband.

Sons married later, and therefore spent longer living with their parents than did their sisters. At this time of their lives they were accumulating resources to enable them to support a family when they eventually married. The shortage of cash in Kikole makes it very difficult for unmarried men to make a living, let alone accumulate wealth to enable them to marry, through paid employment. Many young men, like their sisters, are attracted by the opportunities they believe to be available in towns, and leave the village looking for work. This behaviour confirms Rosenmayr's statement that "The desertion of the old by their adult children is often a result of the powerful forces of economic survival." (Rosenmayr 1988 p37).

Another, smaller group of sons and daughters are sent away from the village by their parents, after completing primary education, to attend secondary or tertiary educational institutions. This further education is provided with the intention of enabling the children to obtain employment, and since there are no opportunities for employment in Kikole, these children do not return to live in the village after finishing their education.

Unmarried sons who remain in the village helped their parents with food and cash crop production, and with physically difficult tasks such as fetching water or firewood. They did not, however, lose sight of their aim of accumulating wealth, even in the challenging economic environment of Kikole. One son, therefore, used his resources to further his own advancement, rather than in the support of his aged parents:

> They've had their chance, and now it's our turn! We should use our money to develop ourselves and forget about them. (18 year old man, grandson of Proscovia Nanyonga EWR=30)

Others were more sympathetic towards the problems of the aged, to the extent of prioritising them above their own, but even they were not always able to help:

> It's not good to forget about old people because sometimes they are weak, and have no money. Then, if you have money in your pocket, you should buy them salt. But sometimes they do have money, and then you can use your own money to solve your own problems. (18 year old man, son of Vincent Katorogo EWR=18)

Whilst some unmarried sons remain in the village, they are not all productive members of their parents' households:

> If they have been helping the old people by getting water or firewood, before they leave for town then the old people miss their help, but some young people are just a burden on their parents and they are glad to see them go to town! (19 year old migrant to Kiguma)

This raises an important issue. Both study participants and I tend to paint a picture along the lines of the "Golden Age" myth, which presents contemporary society in Kikole as having no problems apart from those resulting from modernising changes. In common with all societies, however, relationships between individuals in Kikole were often problematic, and differences or disagreements between friends or family members were highly influential on an individual's access to support. I was told of many occasions when children were said to have abused or misused their parents, or vice versa. These situations might result in the breakdown of the relationship, and the loss of its supportive potential:

> It can be difficult living with an old person. When you go out with your friends and then come back late the old person abuses you and you get annoyed. Then, when you meet a friend who has been to town he persuades you that you should go together and get a good job. When you get to Kampala you forget all about the old person who abused you, and work just for your own needs. Then when you come back, you might find that the old person has reformed; she has been thinking about you, and why you left, and regrets abusing you. She knows that you have punished her for abusing you, and that if she does it again you may go for good! So she doesn't abuse you any more. (16 year old man, grandson of Lilliane Barenzi EWR=27)

> They are wild – uncontrollable. In the past a child would never be rude to its mother, but now they say "I am going! I am going!" and they don't come back. (Emma Kabenge EWR=2)

The provision of support by older, married children

A younger, married, man expressed an understanding of the problems of the older generation:

> If they are weak and can no longer support themselves they get depressed because, although they know what their needs are, they can't satisfy them. An old man may not be able to afford the things he used to have. He might have grown sorghum and brewed beer in the past, so thinks about those things that are now in the past. Eventually he has to give up hope of ever having that pleasure again, but that leaves him feeling unhappy. (29 year old man)

Older children acknowledged that it was their responsibility to support their parents once they became unable to care for themselves:

> If we don't support them nobody else will! If you have a child you must tell him to fetch water for them, and if you don't you must do it yourself, because no-one else will. (40 year old man, son of Julius Kwesige EWR=20)

The traditional division of labour within a household places different obligations on husbands and wives, and the support which a married son can offer his parents is, therefore, qualitatively different from that which a married daughter can provide. A husband is obliged to meet his own family's needs, in particular their cash needs, and to do so before he meets those of his parents. If he is cash poor, as most are, he will be unable to respond to requests for cash support from his parents:

> A son can't help to improve his parents' house until he has done the same to his own. Even though you would like to help your parents, your duty to your family is more important. (25 year old man, son of Noah Sserwadda EWR=12)

Traditionally, a daughter, on her marriage, moves to her husband's land which has been given to him by his father. In many cases this will not be close to her parents' home, and her parents may then have difficulty contacting her when they need help. As a wife, she is responsible for the production of food and the management of her household, and is discouraged from involvement in the cash economy. She is dependent on her husband's goodwill to allow her to support her parents. Such support usually takes one of two forms: giving gifts when making a visit to her parents, or the provision of health care in their home when they are sick:

> For a married woman to support her parents she has to get on well with her husband. She will tell her husband about her parents' problems: "They are sick; what should I do?". If he is sympathetic he will feel sad that his father-in-law is sick and try to find the money to give her to go and see them. If, when she gets there, she finds they are seriously ill, she can return and ask her husband for more support: "I got there and he is nearly dead, what should I do?" Then the son-in-law will try to find the money to send her back. (40 year old woman, daughter of Noah Sserwadda EWR=12)

Married sons, therefore, are expected to provide material or cash support to both their parents and their wives' parents. When sons are poor and unable to meet the needs of both sets of parents, they were said, by one participant, to prioritise those of their wives:

> Married sons think of their own families first, while married daughters give whatever they can to their parents. But these things are given to them by their husbands. I often see my sons packing salt, sugar and soap for their in-laws, but I never get anything from them! (Ibrahim Sabiiti EWR=8)

Social norms today, then, dictate that support in one's old age comes primarily from one's married children, with sons providing cash or material support to parents and daughters obtaining support for parents from husbands. The flow of support from married child to aged parent is therefore seen to be dependent on that child's circumstances.

Parents may expect a return on their investment in their children's education, since their level of education will greatly affect their ability to earn money, particularly in an urban area. In practice, most parents' were prevented, by their poverty, from providing their children with more than a minimal education, the purchase of which served to increase the poverty of their household, but did not allow the children to obtain paid employment. The richest old people were able to provide one or more of their children with a full secondary, and in some cases, a tertiary education, thus equipping them to obtain well-paid employment in urban areas. However, I found not one instance of an educated child supporting an aged parent in Kikole. Parents appear to experience many years of deprivation, during which they provide a succession of children with a poor education, at a high cost to themselves and the other members of their households. For this they receive little, if any return in their dependent years.

Traditionally, a son is given a part of his father's land on which to build a house and rear his family. A married son, living on land close to his father's home, is likely to be more available to help his parents than would be the case if he lived elsewhere, notwithstanding the fact that he also has an obligation to provide for his own family. However, many sons do not provide for their families through the cultivation of their land, but leave their wives and children on the land and go to town to try to earn money. The reasons given for this behaviour vary; parents tend to think that their sons are lazy, and leave the village looking for "easy money" in town, while younger people cite the difficulty in earning enough money in the village to meet the cash needs of their household. Young men said that they did not want to go to town, where they had to work hard for very little money and the cost of living was high, and where they felt vulnerable to exploitation by urban dwellers. If given a choice, they said they would prefer to stay in the village, with their families:

> What forces me to go to town is a poor life in the village. If life here was good I wouldn't go to town. I go there for a job, not looking for a car or a nice house, but just to get money to bring home. I have responsibilities here, like getting salt and sugar, and clothing my children, and I can't stay at home if there is nowhere here to get money! (24 year old man, married with children, son of Noel Katusabe EWR=9)

Whatever underlies the movement of sons and daughters away from the village, the ability of all of them to support their parents is compromised by distance, and by the communication difficulties that exist in rural Uganda. After a man's departure his wife becomes responsible for growing the family food, while he attempts to earn money, through employment in an urban area, to pay the multiple and unavoidable costs of rural life. It may be that his wife helps her husband's parents by providing them with food, or with her labour, or she may allow her children to help their grandparents by fetching them water or firewood, but her ability to provide support will be qualified by her need to support her own children, her limited access to cash, and any obligation she may feel to support her own parents:

> I am married and I don't have any way of earning money. I wake up early in the morning and go and work on my husband's *kibanja*. Then, after digging, I go back and cook for my family. A woman without a family doesn't have problems like mine. I have to keep money so that if my child falls sick I can get treatment, but one without a family can use her money just for her own small problems and sometimes she can save money. If I am a woman without children and I hear that my father is sick I just get in a car and go and see him and take him for treatment. But if I have children I have no money, and I have to find some just to go and see him before he breathes his last. I would like to support him more but have no money to do so. (24 year old woman, daughter-in-law of Noel Sserwadda EWR=12)

The costs of rearing children are identified by this married woman as significantly compromising her well-being. Of all the kinds of support that parents provide their adult children, the most significant is the care they provide for their children's children. As discussed above, dependent grandparents traditionally lived with a grandchild who provided them with domestic help. However, rural-to-urban migration, and, more significantly, the current AIDS epidemic have presented grandparents with large numbers of children in need of a home and their support. The largest proportion of the grandchildren in Kikole who are in the care of their grandparents are AIDS orphans. I shall discuss the impact of the adoption or fostering of grandchildren on their grandparents' well-being in the following chapter, which is concerned with the wide ranging effects of the AIDS epidemic on elders' lives.

The disempowerment of older people is visible in their declining authority over their children as they become older. While younger parents punished or rejected children by whom they were abused or exploited, older, dependent parents were not able to risk alienating their children through such actions or even through criticism, since this might result in the withdrawal of essential support:

> If I hadn't been given supper last night I wouldn't complain to you about it, because if I did, and my children grandchildren found out, they would refuse to cook for me again! (Ester Namukisa EWR=27)

> Thieves are stealing my bananas. They are my grandchildren and my daughters-in-law, but I can't complain, because then they won't give me water which I can't fetch for myself. (George Kabongo EWR=16)

To summarise this qualitative exploration of informal supportive exchange relationships over the lifecourse, I will return to the "life course arena" (Moore 1978) as an analytical tool. An idealised history of supportive relationships throughout the lifecourse can be seen to have three distinct phases. The first is a period when a child or young adult is in receipt of support, the provision of which is a parental obligation, until independence is attained. The second is a period during which exchange relationships with friends, voluntarily entered into, are the major source of support. Thirdly, a period of declining productivity develops, when the reciprocal obligations of supportive friendships can no longer be sustained, and when children are obliged themselves to support their parents.

The boundaries between these phases are not defined, nor are the phases mutually exclusive. The ability to obtain support from friendships is dependent to some degree on their number, and young adults who are in the early stages of the process of accumulating these relationships are likely to have more cause to call on a parent for help than older adults who have many such relationships to draw upon. Parents can thus expect to continue to support their children in the years after they have left home, as they are establishing their own supportive friendships. Further, more so for aged men than for aged women, it is very likely that they will be supporting both young and adult children, and possibly grandchildren, at the same time. Later in life, the withdrawal from friendships and the return to family support is likely to be a gradual process rather than a single event. At this time, the adult children upon whom aged parents are calling for support are not only caring for their own children, but are also in the process of establishing their own network of supportive friendships. Any support they give to their parents cannot be invested in these reciprocal friendships, and will therefore reduce the extent of the adult child's support network. Moreover, since this support is offered in return for that previously supplied by parents, these parents may not feel any obligation to reciprocate. Adult children, who acknowledge that they have to prioritise the welfare of their own family above that of their parents, may be reluctant to support their parents if to do so puts at risk the future well-being of their family. Figure 6.1 illustrates the shift in sources of support over the idealised lifecourse:

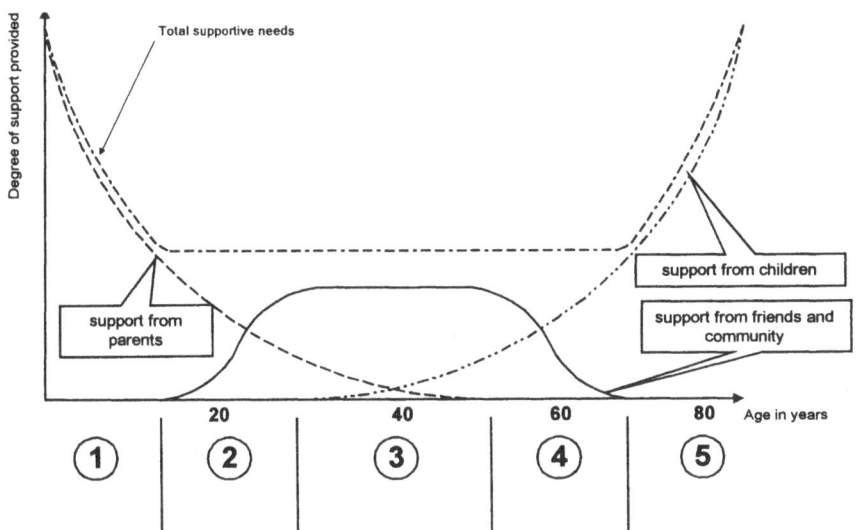

Figure 6.1   The receipt of support throughout the lifecourse

This figure should be considered as indicative rather than definitive. It is a representation of the availability rather than the supply of support. The amount of support received at any moment will be related both to the needs of the recipient and the circumstances and resources of its providers. The figure addresses only the major sources of support, small amounts of which may be received from other sources. The figure divides the lifecourse into five segments:

1. The period from birth until the time, during adolescence, that the individual begins to develop supportive friendships. During this period all support comes from the parents.

2. The period from adolescence until one begins to offer support to one's parents. During this time friendships grow in number, and support is obtained from them and from one's parents, who may still be able to offer support in some areas, whilst being in need of it in others. The net flow of support between parent and child remains towards the child.

3. The period during which most support comes from reciprocal relationships with friends.

4. The period during which one's abilities start to decline and one becomes less able to maintain reciprocal friendships. Support obtained from these relationships therefore declines, and one becomes increasingly dependent on one's children.

5. The period after withdrawal from reciprocal relationships with friends during one is entirely which entirely reliant upon the support of one's children.

Marriage as a supportive relationship

The marital relationship is qualitatively different from those so far described, since it is, potentially, a continuing source of support for both husband and wife during the periods in which they are raising their children, and after these children have become independent, and into their old age, until the end of the relationship, whether due to separation or the death of one partner.

Since the partners in a marriage have separate roles in the management of their household, the support sought by each partner reflects their role. Thus, since a husband is expected to meet the financial needs of his household, his friendships are developed with other men, and the support he obtains from these relationships helps him in this task. A wife's duties are centred on the provision of food, and she develops relationships which will increase her ability to do this. Her friends may give her surplus food from their land, or, if they have control of land, they allow her to grow food on an unused portion of their land.

The following quotation, describing the activities of a co-operative group who dig together, for money, illustrates the separation of the roles of husband and wife:

> There are three of us now. Me, the wife of Deogratius, who lives nearby, and my deceased brother's son, who lives next door. We all dig for three days, one day on each of our *kibanjas*, and on the fourth we go and work for money. On the fifth day we resume working for each other. We share the money equally, immediately after finishing the job. If one of us has a problem and can't dig then we don't dig together that day, and each of us works in our field. Today, for example we aren't working because the wife of Deogratius has visitors – her daughter and her daughter's husband are coming to sympathise for the loss of a relative. (Rebecca Nambiru EWR=22)

Deogratius was a regular employer of Solomon Nsamba (EWR=13), Rebecca Nambiru's neighbour. Deogratius had cash available to pay Solomon to improve his cash crop production, while his wife was working for money to meet her own needs, and, through co-operative labouring, to increase the food production of her husband's land. It is not Deogratius's role to use his cash surplus to increase food production, nor his wife's role to work in her husband's coffee plantation.

When their capabilities start to decline, the boundaries of spouses' roles become less distinct, and they may share their responsibilities. In old age, a spouse, male or female, is unlikely to be able to generate support from outside the household, but his or her presence within the household at this time may reduce the need to seek support elsewhere in coping with declining ability. Therefore, the loss of a spouse will force the remaining partner to call on his or her existing relationships to provide the domestic assistance that has been lost upon the death of the partner:

> An old person who is alone faces more problems than a couple, because when there are two one can go and get firewood or harvest food from the plantation while the other is doing something else. But the person who is alone has to do everything for themselves. (41 year old woman)

Writing of Bangladesh, Rahman (1997) reports a higher mortality risk for the unmarried compared with their married contemporaries. He also asserts that aged women, as a result of their lower economic status, are more vulnerable to the negative effects of widowhood. Men, through their greater control of resources and their ability to earn their own living can respond more effectively to the problems that arise from widowhood, and are less dependent on others for support at this time than are women.

In Kikole, the separation of their household roles presents men and women with differing problems when they are widowed and their supportive relationships may not be able to respond appropriately to their needs. Widowed men had domestic problems, while women had difficulties managing their deceased husband's farm:

> There are problems for men living alone because all their lives many things have been done for them. Washing is done by someone, food is brought by someone, and if his wife dies there is no-one to do these things. But a woman can do these things. Even if it takes two hours for her to wash the clothes she can do it. It is not the same for a man. (34 year old man)

> Widows are usually poorer than widowers because they don't know about farming. Women who have been independent can be OK, but those who have never had full control of their lives because their husbands planned everything for them have problems. They can't take over his job because they don't know how. (31 year old man)

The first of these comments contradicts Rahman's (1997) assertion that men cope with widowhood better than do women, while the second supports it. The key issues are the availability of support to the surviving partner after the other's death, and the material wealth of the husband. If the husband is poor, then when he becomes a widow he may not be able to support himself. He will become, as would his widow following his own death, dependent on supportive relationships. A wealthy husband, however, can use his resources to support himself after his wife's death, whereas a widow who, during her marriage, had been dependent on her husband, may not have access to his wealth, or have the skills to exploit it, after his death.

Isolation is a problem for a married woman. A husband is required to travel, in the village or further afield, which gives him many opportunities to meet other men with whom he might establish friendships. A married woman's domestic responsibilities, however, restrict her opportunities to leave the home and establish supportive relationships with other women. Further, if, like most older married women, she is not involved in the cash economy, she will not be able to enter into supportive relationships which are based on the loan and repayment of cash, and will be dependent on her husband for all her cash needs. If she becomes a widow her friendships, which are centred on the mutual fulfilment of a wife's duties, will be unable to supply her household with the support that her husband was able to supply through his cash-oriented relationships. At this time she will therefore be in need of support from her children, and as has been indicated above, her children will not always be able to provide for their widowed mother.

Not all unmarried women are in this position through becoming widows. Four had chosen this status, preferring not to be married, and another had been expelled by her husband from his home after producing her children, and had decided to remain unmarried. It is interesting, at this point, to contrast the lives of two groups of these voluntarily unmarried women. In the first group, two of these women had made the choice to remain single when they were young, one having had three children with three men, with none of whom she cohabited at any time, while the other had left her husband when her children were young, and reared them herself. Both of these women were educated and had lived in town, accumulating wealth in their own right through their respective jobs as a midwife and a teacher, and educating their children. They moved to Kikole after their children had matured, left their mother's homes and obtained jobs in urban areas. In Kikole, both women now own large *kibanjas*, and grow and sell large amounts of coffee. One of them continues her professional practice as a midwife. Their emic wealth rankings are two and six.

One of the other three women decided to remain unmarried after leaving an abusive husband, and returned to the village to live on land given to her by her brother. The second left a promiscuous husband and moved to the home of an adult, unmarried daughter, and the third was sent back to her father's land after many years of marriage, following a dispute with her husband. All three were minimally educated, and none had accumulated wealth during their marriage, or been able to do so since returning to the village. They are now impoverished, with wealth rankings of 22, 25 and 26.

The differences in the circumstances of these two groups of women illustrate the relevance of previous lifecourse decisions and events to present well-being. The two who had purchased land were following Baganda women's "obsession" with land ownership, identified by Obbo (1986), as an insurance against poverty in their old age. However, land purchase may prove not to have been purely beneficial for the wealthier two women, since neither of them, having moved to the village in their middle age, have any extended family members living nearby who can be called on for support. They are thus reliant on their own, albeit relatively plentiful, resources to respond to any difficulties they face.

Today's young women who choose to go to town to maintain their independence, are repeating the behaviour of these few aged women in Kikole who have chosen to remain unmarried in order to establish their economic independence. Southall and Gutkind (1957), working in Kampala at about the time that these women were making their decision, noted that there were many women living unmarried lives in Kampala. That this was a successful strategy can be judged from the fact that the two wealthiest old women in the study had both made this choice, and were wealthier than most of the old men in the study. However, they were very much the exception, as the other 15 old women had married, some being still married and others now widowed.

Supportive relationship with kin other than children

There were few reports of exchanges taking place with other kin, but this has to be seen in the context of the history of settlement of the village, as has been discussed in Chapter 3, and which has resulted in few aged adults having kin other than their children living in Kikole. Two sisters lived together, and described the help they gave each other:

> It's like living with a grandchild. We help each other and don't quarrel. What would happen if we were alone and one of us got sick and died? Who would tell other people? (Polly Katate EWR=7)

There were two groups of aged brothers in Kikole. They socialised together, and helped each other with occasional loans of money. Although these relationships were stronger than a typical friendship, the ageing brothers' declining participation in the village economy rendered their mutually supportive relationships less significant then those they maintained with their children.

Priorities and decision making in the utilisation of informal supportive relationships

The time at which an adult child becomes independent is, potentially, the start of a time of improving well-being for the parents, who are now able to shift their investment of resources into the establishment of supportive friendships. Parents will have to wait until their old age before they receive reciprocal support from their children, whereas reciprocal support from a friendship will be offered within a much shorter timeframe. Further, there is the possibility that an adult child will not provide support to a parent, due to distance, reluctance, poverty, illness or death, or living too far away. In this case any investment in that child is lost, and when this becomes apparent, the dependent old person has no opportunity to "reinvest", by having more children in the hope of a better "return". In contrast, friendships that fail to deliver reciprocal support can, since they are conducted always with an eye on this possibility, be ended before many resources have been invested and lost, and while there remains the opportunity to establish new, more productive relationships. At this time, older men are advantaged in comparison to older women, since, if they have resources, they are able to expel an ageing wife and invest them in the acquisition of a younger wife who will care for them as they become dependent. Older married or widowed women, however, since they rarely own land or sufficient resources, are very unlikely to have this option open to them. Even if they did, the experiences of the unmarried women, described above, suggest that they would prefer to remain unmarried.

Let us assume, for the moment only, that decision making regarding the support of one's children is always conducted with the aim of profit maximisation. Within the word "profit", here, I am including both monetary and material goods, but not psycho-social well-being. A father, when faced with a request for support from an adult and theoretically independent son, has a number of factors to take into account when considering how to respond. The child is already obligated to support his father when, at some point in the future, he becomes dependent, and not before this time. Any additional support given to this son is, therefore, unlikely to result in his reciprocating, years later, with the provision of additional support to his ageing father. The nature of their reciprocal relationship does not allow for the keeping of a tally to ensure that a balance is achieved in the amount of support exchanged between the birth of the child and the death of his parents. If, however, the father invested his resources in his friendships, they would theoretically produce reciprocal support in direct relation to the amount invested, and therefore be more productive than an investment in one's children.

Thus, it becomes apparent that an older adult, whose children are independent, and who has a limited amount of resources to invest in the reciprocal relationships that comprise his social and economic security system, would maximise the return on his investment if he invested solely in his friendships, and ignored the needs of his children. Any further investment in his children would fail to produce any increase in the support they already feel obligated to provide, and also reduce the amount the parent has to invest in his productive friendships.

In practice, of course, decisions of this type are not made on such a hard-headed and detached basis. Non-economic factors such as feelings of obligation and affection influence relationships with both children and friends, and both psycho-social well-being and profit maximisation play a role in the negotiation of supportive exchanges. The obligation to support a son or daughter is a powerful one, particularly if the request is for support needed to meet the needs of a son's children, towards whom their grandfather will also feel a responsibility, since they are also of his clan. The strength of this obligation will usually force the father to support his child, if he is able, even though to do so will reduce his ability to establish productive friendships. Here we have confirmation of my earlier assertion that the Marxian analysis of the peasant economy at the population level cannot be extended to analyse individual economic decision making, since decisions are made with reference to non-economic criteria.

The nature and extent of support provided by adult children

Nydegger (1983) notes the lack of evidence that multi-generational households were the norm in pre-industrial societies, where migration of children away from the family residence was not unusual. Evidence to the contrary suggests that in many African cultures older people did indeed live in close proximity to their extended families, although not always in the same building as their younger kin. Bennett and Mugalula-Mukiibi (1967) noted that old people living alone in Buganda were doing so as a result of fear of *obukko*, but that they were not socially isolated, since they were living near relatives and were visited by them daily.

Nydegger (1983) also asserts that the separation, as a result of migration, of adult child and aged parent need not present problems for the latter if communication between them is not problematic and if the child could travel quickly to its parents' assistance when needed. However, Kikole has no telephone or radio link with the wider world, no reliable public transport system, and most people have no money with which to hire a vehicle or to pay to send a messenger to town. The mail service is unreliable, and illiteracy limits its effectiveness. The most common channel of communication is an informal system in which a written message, addressed to a distant individual, is carried by travellers, free of charge. The message will be passed from hand to hand until, after an indeterminate period, it reaches its addressee. Several of my conversations with elders were interrupted by the arrival of a stranger carrying such a message, which in every case bore news of the death of a relative in Kampala. Poor communications created difficulties for those in need of help:

> My daughter went to Kampala to meet her sisters who live there, but she hasn't come back. Now I am ill and there is no-one here to send to tell that one of them should come back and care for me. (Noel Katusabe EWR=9)

It can be seen, then, that an older person facing a crisis in Kikole will have problems communicating this difficulty to children living more than a short distance away, and since help in one's old age comes primarily from one's children, that those old people without children living relatively nearby will be particularly vulnerable if they, for example, meet with an accident or are taken ill. One unmarried young man, still living with his parents, explained how he was of more help to his parents than were their more distant children:

> Sons help more than daughters because they live nearer to their parents and will be approached for help first for that reason....My mother has many daughters who have all gone away after getting married. Some are in Kampala, and others in Jinja, while I am at home, helping our parents. My sisters have more money than me, but they only come to see our parents once or twice a year. When they are sick, they ask me for advice, because they wouldn't know where to find their daughters in Kampala. (21 year old man)

The following paragraphs contain an analysis of the type of support offered by children, and confirm that this support is indeed largely offered by children living nearby. To learn the distance between participants' homes and those of their children, I asked parents to tell me where these children lived. The answer to the question: "Where does [name of a son or daughter] live?" was usually: "In [name of a village], just past [name of another resident of that village]'s house", or a similarly imprecise reply, which did not facilitate a calculation of the distance between their home and that of their children. The definition of "nearby" in the following discussion, is, therefore: "Living in the same house, the same village or in a neighbouring village." In practice this means a distance of no more than four or five kilometres, a distance which I have taken to be the maximum which a child would be prepared to travel regularly, on foot, to visit his or her parents. This is also close enough to allow a message to be sent for urgent help when it is needed. Both these factors are important to the maintenance of the well-being. They allow for regular social visits and gifts of food or labour to be made, and, should the elder be taken ill, for a child to be contacted and to come and provide care or take him or her to a health care facility should that be necessary.

Table 6.1 shows this to be an important factor in the well-being of aged people in Kikole. It shows that the heads of those households with emic wealth rankings in the lowest third of the range have fewer children living nearby than other households:

Table 6.1   Emic wealth ranking of heads of households aged 60+ and mean number of children living nearby

Emic wealth ranking of head of household		Mean number of nearby children
1-10	(n=10)	2.1
11-20	(n=8)	2.5
21-30	(n=8)	1.0
All households	(n=26)	1.9

Since wealthy people usually have more land than poor people, they are likely more often to be able to give their children, usually their sons, sufficient land from which to make a living, and therefore to have more of their children remain resident nearby. The content of Table 6.1, above, should therefore not be a surprise.

Table 6.2 shows that women, especially unmarried non-Baganda women, have fewer adult children living nearby than the rest of the population. To understand why this should be so, it is necessary to explore the reproductive history of aged men and women. The wife of an aged man is usually younger than her husband, and these men therefore have young children in their households much later in their lives than do aged women. Thus, even though old men's adult children may leave the village, their younger children continue to mature, and then to become available to support their parents. Elder women, on the other hand, produced their last children at least fifteen years ago, and in the intervening years their children have lived hazardous lives, in unhealthy environments, and during periods of civil war. Some have died and others have vanished, their fate unknown. These women can only hope that their children survive to care for them in their old age, and do not have the security of the knowledge that they can produce more children, late in their lives. The higher numbers of adult children living nearby is therefore, in part at least, a result of having more living children overall. This is found to be the case, and Table 6.3 shows that old men have more living children than old women.

Table 6.2   Mean number of children living nearby by marital status and ethnic origin of parent aged 60+

	Mean number of adult children living nearby by ethnic origin of parent aged 60+		Total
	Baganda	Non-Baganda	
Married men	3.0 (n=4)	2.4 (n=8)	2.6
Unmarried men	2.0 (n=1)	0	2.0
Married women	1.0 (n=1)	2.5 (n=2)	2.0
Unmarried women	1.7 (n=6)	0.8 (n=8)	1.1
Total	2.1 (n=12)	1.7 (n=18)	1.8

Since only married aged men still have young children, Table 6.3 changes little if only adult children are considered, rather than all children. Table 6.4 shows that all unmarried men's and women's children are already adult.

Table 6.3   Mean number of surviving children by marital status and ethnic origin of parent aged 60+

	Mean number of surviving children by cthnic origin of parent aged 60		Total
	Baganda	Non-Baganda	
Married men	17.7 (n=3*)	6.6 (n=8)	9.6
Unmarried men	3.0 (n=1)	0	3.0
Married women	4.0 (n=1)	5.0 (n=2)	4.7
Unmarried women	3.3 (n=6)	2.0 (n=8)	2.6
Total	7.3 (n=11)	4.4 (n=18)	5.5

* Two of these three are polygamous, hence their large number of living children. None of the non-Baganda men are polygamous. This figure excludes one of the households included in the previous table; the polygamous man had another household a long way from Kikole and it proved impossible to discover exactly how many children he had living there

Table 6.4   **Mean number of surviving adult children by marital status and ethnic origin of parent aged 60+**

	Mean number of surviving adult children, by ethnic origin of parent aged 60+		Total
	Baganda	Non-Baganda	
Married men	12.7 (n=3)	4.8 (n=8)	7.0
Unmarried men	3.0 (n=1)	0	3.0
Married women	4.0 (n=1)	5.0 (n=2)	4.7
Unmarried women	3.3 (n=6)	2.0 (n=8)	2.6
Total	5.9 (n=11)	3.6 (n=18)	4.5

These three tables, when considered together, indicate that although the married, aged, Baganda men have more living adult children, a lower proportion of these live nearby than do the adult children of the other old people in Kikole. In absolute terms, however, they still have more adult children living nearby than other aged people. These Baganda households are amongst the wealthiest in the village and have educated many of their children, both sons and daughters, to secondary or tertiary level. Kikole provides no opportunities to utilise one's education. These children, although they are not typical of young people in the village in their educational status, move, along with other, less well-educated children, to urban areas in search of work. It seems that the advantage that they may experience as a result of the relative wealth of their parents is not sufficient to persuade them to stay in Kikole.

While support such as a gift of *matooke* or labour, can be provided at no financial cost, other forms of assistance, such as a gift of meat, or the payment of health care costs, may only be provided by those with sufficient funds to purchase them. There were no examples, in Kikole, of aged parents being supported by their wealthy, urban dwelling children. One may therefore hypothesise that, since support comes primarily from children who live nearby, and who experience varying degrees of impoverishment, the support they provide will be largely in the form of labour, or of material goods that involve the donor in no financial expenditure. To explore this point further, I will discuss elders' households' access to four types of support, in relation to the location of their children's residence. Two of these, the provision of health care and the supply of meat, require financial expenditure on the part of the child, while the others, the supply of agricultural labour and of *matooke*, do not.

Table 6.5 shows the relationship between the ability of a household always to obtain health care when it is required, sex of the head of household, emic wealth rank and the number of adult children living nearby. It indicates that the households of the participants who were always able to obtain health care when in need were significantly wealthier, but had only slightly more adult children nearby, than did the households of those who could not always obtain this care. Thus, the ability to obtain care was more dependent on household wealth then it was on the number of adult children living nearby:

Table 6.5   Household use of health services, emic wealth ranking of head of household aged 60+, and the presence of nearby adult children

Able always to obtain health care	Mean emic wealth ranking of head of household*	Mean number of adult children living nearby**
Yes (n=7)	6.1	2.1
No (n=19)	17.8	1.8
Total (n=26)	14.7	1.9

$$* f = 11.889, p<0.005 \text{ (ANOVA)}$$
$$** f = 0.197, p=0.66 \text{ (ANOVA)}$$

"Famine" occurs in Kikole because, although *matooke* ripens all the year round, it was not always available due to the declining yields in the village. Very dry weather would also slow the ripening of the fruit. Even the wealthiest people could not force their land to produce *matooke*. When it was not available most people consumed the staple that they did have ready for consumption, usually the less desirable and palatable sweet potatoes or cassava, rather then using scarce cash to buy *matooke*. Thus, in spite of its desirability, consumption of *matooke* was not associated with household wealth, as Table 6.6 shows. This table shows that aged parents who, throughout the study period, consumed primarily *matooke*, had more adult children living nearby, than did parents who consumed primarily cassava or sweet potatoes, suggesting that the nearby children were supporting their parents through the provision of *matooke*.

Table 6.7 shows that since meat is available only to those who have money to buy it, consumption of this food is significantly related to the wealth of the household. The presence of nearby adult children, who are unlikely to purchase meat for their parents, is not related to meat consumption.

Table 6.6   Most frequently consumed staple of head of household aged 60+, mean emic wealth rank and location of adult children

Staple most frequently consumed		Mean emic wealth rank of head of household*	Mean number of adult children living nearby**
Matooke	(n=13)	14.6	2.3
Cassava or sweet potatoes	(n=13)	14.8	1.5
Total	(n=26)	14.7	1.9

* f= 0.002, p=0.96 (ANOVA)
** f= 1.509, p=0.12 (ANOVA)

Table 6.7   Household consumption of meat, mean emic wealth rank of head of household aged 60+ and location of adult children

Meat eaten at least once a month		Mean emic wealth rank of head of household*	Mean number of adult children living nearby**
Yes	(n=12)	10.1	2.0
No	(n=14)	18.6	1.7
Total	(n=26)	14.7	1.9

* f=6.61 p<0.05 (ANOVA)
** f=0.272 p=0.6 (ANOVA)

Those with declining ability have increasing difficulty cultivating their land. If unaddressed, this will inevitably lead to a decrease in their food or cash crop production. This, in turn, is likely to have a negative impact on their well-being. The only solutions available to those who are unable to work as long or as hard as in the past, are to employ casual labour or to persuade a relative or a friend to provide this labour as a favour. In Kikole, aged people did not work on each others' land, except in the formal, co-operative situations that I describe in the following section of this chapter. Nor did married children, even those living nearby, assist their parents by working on their land, since their first duty was to secure the well-being of their own household. Rather than provide labour they provided their parents with the surplus produce of their land, such as *matooke*. The only adult children who helped with the cultivation of their parents' land were co-resident children: unmarried sons or daughters who were living in their parents' household, and contributing some or all of their labour towards the household economy.

We can now see the pattern of assistance that children are able to offer their parents. While they are resident, they can help with many of the problems their parents face, but once they move away their ability to help is reduced through the acquisition of a family and the primary obligation to ensure its welfare. If they live nearby they will be able to support their parents through social visits, and may be able to provide material help during a crisis. Their poverty, however, usually prevents them providing monetary assistance. Distant children do not send support to their parents, although they may bring presents when they visit. Very often, however, they do not, in which case the parent has to accept the costs of entertaining his or her guests for as long as they wish to stay, and the visit involves a net cost rather than net benefit to the parent. As indicated above, all parents are vulnerable to this event, and villagers consider it appropriate for friends to help each other should this occur.

Table 6.8   Mean emic wealth ranking by emic age ranking of all adults
aged 60+

Emic age ranking	Mean emic wealth ranking
1-10	11.6
11-20	12.0
21-30	22.7

$p < 0.05$ (Kruskal-Wallis)

A possible confounding factor in this analysis is the relative wealth of the nearby adult children, since it is possible – even probable – that the children of relatively wealthy parents will themselves be wealthier than the children of relatively poor parents. But since I have shown that nearby children neither support their parents with gifts of meat nor health care, both of which are monetary expenses there is no empirical data to indicate that wealthy nearby adult children are of more benefit to their parents than are their poorer contemporaries. Of course, the lack of support reported to be supplied to wealthy parents may simply be a reflection of their lack of need. However, this assumes that they are able to provide for themselves. Whilst they are relatively young this may be true, but an old and frail person is likely to need support whether they are rich or poor. One might, therefore, hypothesise that a wealthy person who has wealthy children living nearby will not experience a decline in circumstances as he or she ages. However, the relationship between emic wealth rankings and emic age rankings, shown in Table 6.8 indicates that the oldest participants were also the poorest, suggesting that even if wealthy children are living nearby, they do not prevent a decline in their parents' circumstances as their ages increase.

Childless elders

Since adult children are the primary source of support for elders in Kikole, childless elders can be expected to be more impoverished than others. There was only one childless elder in the village and she lived with her younger, but also aged, sister. The sister, whose extremely wealthy son lived nearby, was given an emic wealth rank of seven, whereas the older, childless, sister, had a low emic wealth rank of 24. Since the women lived, worked and ate together, and shared resources, this difference suggests not only that living children are equated to wealth, but that the wealth they represent is manifested in the potential support they can provide their parents. It also implies that those without children will find it difficult to obtain support from elsewhere:

> A childless man will not be supported because people are poor. There is a proverb *"Obwavu tebukumanyisa akwagala"* – it means "A poor man can't show you his love" In other words; even if we want to help, we can't, because we are poor. (67 year old man)

> We don't help each other! Your neighbour will say he loves you, but he won't help you when you have a problem. The only people that will help you when you are dying from hunger is your son or your daughter. (Irene Mutondo EWR=30)

The woman who made the second of these statements did, in fact, have children living nearby. Apart from confirming their importance she also raises the issue of seasonality; in a poor season everyone is likely to be short of food or money, and at such a time blood ties are more significant than those between friends and neighbours. Thus, while childless people may be given help in times of relative plenty, they will be abandoned when resources are in short supply:

> All the flowers were knocked off our beans during the hailstorm, and the cassava was damaged, so famine is coming. Those who have the strength will work for others to buy flour, but those of us who are weak will starve to death! (Sulaiman Kagire EWR=10)

Formal exchange relationships

Short term formal exchange relationships

These are usually formed in response to the unexpected need for a large amount of money by one party, eg. to pay for taxes, school fees, a brideprice, or health care costs. If the amount needed is more than can be obtained through calling on one's informal exchange relationships with friends or family members, one has either to enter into a formal relationship with someone who does not fall into either of these categories, or formalise an existing informal relationship. The latter is unlikely to be possible in the case of friendships, since as I have discussed earlier, these are usually formed between individuals of similar economic standing and, therefore, if one party does not have sufficient resources to address the problem then the other party is likely to be in the same situation. Thus, new relationships are entered into to address problems of this kind, and money is borrowed from people who one may not know well, and whom one may not trust in the same way that one trusts the friends with whom one has informal supportive relationships. The urgency of the problem may force the debtor to accept disadvantageous terms, while the creditor may charge a fee or interest on the loan. A date for repayment is usually fixed, confirmed by a written agreement to this effect, signed by both parties and witnessed by others. This is so even though it is likely that one or more of the parties will be illiterate.

Seasonal variation and unpredictability of future income, health problems, or unanticipated financial needs may cause a poor person who has borrowed a large sum of money to experience difficulty in repaying the debt. Borrowing money in this way was, therefore, a last resort; I recorded only one instance during the study period. One participant, whose use of health care services was described in Chapter 5, had borrowed money to pay for this care, and repayment of this debt caused him problems:

> I borrowed money when I fell out of a tree and needed treatment. The trouble is that we agreed that I would repay in coffee rather than cash, and at that time the value of fresh coffee was Sh2000 per tin, but now it is Sh5000. I borrowed Sh30,000, the equivalent of fifteen tins at Sh2000, and so far I have given them ten tins, which makes Sh50,000 at today's prices! But they came and told me to pay the rest immediately or they would take me to the police. So the coffee you can see here is going to be sold at Sh2000 even though the real price today is Sh5000!.....I have asked the chiefs on the LC to help me; I don't mind paying as these men helped me, but they are asking for too much!

The village council was asked to arbitrate in this dispute. By the time their decision was made, he had repaid another three tins, and remained with a debt of two tins of coffee. The council, which has only advisory powers, asked him to repay these and asked his creditors to give him a further Sh10,000. Both parties accepted this arrangement, and the old man had thus received Sh40,000 in return for coffee with a market value of Sh75,000. Although this was the only example I found during the study, the fact that all parties appeared to think this was an equitable outcome suggests that loans in Kikole were routinely subject to high interest rates or other charges.

Long-term formal exchange relationships

Mair (1934) notes that all relatives and friends who came to offer condolences to the bereaved were expected to bring a gift – a bark cloth in which to wrap the body for burial, or money with which to buy this. Mair did not mention the obligatory and reciprocal nature of these gifts, but today, all members of the community are expected to contribute towards the burial costs incurred by one of their number. A death, and the time of the burial, is announced by a man on a bicycle, through a megaphone, and the ceremony is a relatively brief event which all villagers were expected to attend. The costs of a burial can, however, be high, since family members may come from elsewhere and stay several days, requiring the bereaved household to provide food and drink for them. Every household is expected to pay an *mugogo* ["condolence fee"] towards these costs. Each contribution, which may be as little as Sh100-200 in some cases, is received formally, and publicly, by a senior member of the family, and recorded in a book, either by this person, or if he is illiterate, by another, literate, member of the family. These records are then used, by each household, on the occasions of future burials to ensure that contributions are repaid in equal measure. The reciprocal obligations of friendship were considered when deciding how much to give:

> If you are a close relative you may take a bark cloth for the body to be wrapped in, or if the burial is in the family of a close friend and you can be sure he will repay what you give you may give a lot, but otherwise you just give what you have in your pocket, maybe Sh300 or Sh500. (Sulaiman Kagire EWR=10)

This system was scrupulously observed by almost all elders, since failure to do so was seen as shameful:

> It is our custom that we have to pay *mugogo* at each burial, but some people don't give because of poverty. That has caused a lot of shame for old people. If I don't have any money I don't go to burials. Imagine! Going to sympathise but failing to pay the condolence fee! (Noah Sserwadda EWR=12)

Not only is it shameful to fail to contribute to the costs of these events, it also compromises one's future well-being. By making contributions towards the costs of one's friends' and neighbours' ceremonies, in effect one makes a series of deposits in a savings account, the accumulated funds of which can be withdrawn when one is faced with the costs of a burial of a family member. If one does not make deposits, then there are no funds for withdrawal when one is in need. Therefore it is considered not only shameful, but economically unwise to fail to contribute to this fund:

> You give to the person who has the expense of burying someone. You would give to that person even if you didn't know the person who has died. The amount your relatives will receive when you die depends on the amount they give to others and not on the amount you have given. (28 year old man)

Labour could be donated by those who had no money. Men helped with the erection of the temporary shelters that were provided for the guests, and served them with food that had been prepared by the women of the village. If an elder was unable, through poverty, to donate money, nor to contribute labour due to frailty, it was considered acceptable to send a younger member of their household in their place, but if no effort was made to contribute then one could expect no help in return. One woman, whose house was isolated, who took no part in the life of the community, and who had no children living nearby, decided not to worry about her burial, and thus save the money she might have invested as condolence fees:

> I have stopped going to burials. I can go a whole month without anyone coming past my house, so they'll only know I'm dead by the bad smell. They may not even find me before I've been eaten by dogs! If people want to come to bury me they can, but I don't care. I won't know anything about it! (Christine Nabagareka EWR=23)

An *orumbe*, at which the heir to a deceased person was identified, and the will read, was usually held some weeks or months after the death, the interval being determined largely by the time the family needed to raise the money necessary to pay for the ceremony. The ceremony is intended to honour the spirit of the dead person and as a formal ending to the mourning period. It is important, if possible, to spend a lot of money on a lavish event, since a dissatisfied ancestral spirit can cause illness or misfortune for the family. To this end due ceremonial has to be observed. This may demand the presence of a priest, who will require payment for his services. All guests have to be fed well, preferably with meat. Invitations are extended to distant family members and to friends living away from the village with whom those organising the *orumbe* had reciprocal relationships. Additional invitations are sent to those who are seen as possible sources of significant amounts of financial help. Here, once again, the economic stratification of supportive relationships is evident:

> Poor people aren't formally invited to ceremonies because they can't help with problems, but now that I am not so poor as I used to be I am invited to more ceremonies because they know I can help. And I go because I know they may help me in the future. (31 year old man)

Guests arrive on the day of the event, with contributions of money, food or drink. Those who attend an *orumbe* are expected to contribute towards its costs, although there is less obligation on the entire community to contribute, than there is in the case of a burial. This was said to reflect the urgent nature of a burial, which had to take place soon after death, and for Moslems the day after at the latest, whereas an *orumbe* could be delayed until sufficient money had been raised to allow an appropriate ceremony to be held. Those who chose to help would make a gift of cash, food, or firewood for cooking the large amounts of food served to guests. Records were kept in the same way as for a burial, and repayments in equal measure made when others were organising an *orumbe* for a member of their own family.

Other formal co-operative enterprises

The contributions made by villagers towards the ceremonial costs of their friends or neighbours are a rare example of formal economic co-operation within Kikole.

Other co-operative ventures had repeatedly failed due to theft of funds by their officers, generating a widespread cynicism among the study participants, and their reluctance to become involved in such an organisation. *Munno mu kabi* ["Friends in times of danger"], an umbrella term for a variety of community-based organisations formed with the objective of helping their members deal with specific problems, had been widespread at one time, but are rarely found today. Only one other organisation was functioning, and the fragility of its existence can be judged from an old woman's experience of membership, described in Box 6.1.

Box 6.1 *"Agaria waamu"* ["Being together"]

I have joined a small association called *"agaria waamu"* ["being together"]. Every member is helped when they have a problem, because we each contribute a thousand shillings to help that person. At first each of us paid monthly, on the thirtieth, but then the secretary for finance stole the money so we stopped paying. Next month we are going to start paying again, and each one of us will give Sh500 when a member has a lost a relative, and for a wedding or a funeral rite ceremony it will be Sh1000.

   It started last April, but I didn't know about it then. I joined after I had buried my brother last year, when I heard people talking about it at the burial. I would have joined at that time but my daughter was also ill, and I had no money to contribute. She died in December and I joined in January, so I have been a member for nine months. There are eighteen members, and five of us are from this village. Most are younger than me, but there are two other old women who are members. The membership is closed now and no more people may join. This is because it isn't easy to keep control when there are many members. I don't know why that is, but in the past, when we had a lot of money and we organised a feast, people quarrelled and someone stole the money.

   The son of my brother who died is getting married soon, and the members will help me with Sh10,000 towards the cost of the wedding.

Similar organisations established in the past had foundered following disagreements among their members. Once again, stratification along economic lines was a factor:

> Rich people would only contribute when it was for a rich person, and eventually it divided into two – a group for rich and another for poor people. (Henry Ssewanyama EWR=15)

Another man's complaints revealed one of the few causes of tribal tension in Kikole:

> My wife and I joined *munno mu kabi* years ago, but we weren't treated well at burials; not with the respect that members of that organisation should get. Some members were thought to be more important than others, and got special treatment, better seating and

food, but we didn't. The problem is tribes. The group is for the whole community, but it's the Baganda who get the good treatment and not us. They wanted the tribes to sit separately, and it hurt us in our hearts because we had all contributed in the same way. So, we didn't feel valued by the group, and we left it. We've never had that kind of problem with weddings or funeral rites. (Sulaiman Kagire, an old man from Burundi EWR=10)

Corruption had caused the demise of other co-operative ventures:

We had a women's group that made handicrafts; we used to make things and sell them, but the leaders took all the money. Then about three years ago there was another club, a big one, organised at the sub-county level. We planned to keep silkworms, and chickens and bees, and we all contributed some money, but our plans were never implemented because all the money was taken by the leaders again. (Grace Nanteza EWR=6)

Smaller scale, co-operative, digging ventures were more successful. Two women worked together, sharing the proceeds of their labour, although the continual shortage of cash in the village made such work hard to find:

We look for a piece of work, negotiate a price and when we have completed it we share the money. Today we have been working for a woman, digging a piece of land for Sh2000. We expect to finish it in two days, perhaps tomorrow, and get the money. I am not expecting any other income as I can't find work. They say they can't afford to employ us. (Rebecca Nambiru EWR=22)

Supportive exchanges with the state

As noted in Chapter 4, all adult men and some women are taxed by the government each year. Since there were no defined benefits to be gained from making these payments, first impressions suggested that the government provided rural people with very little in return for their taxes. There were no free education or health care services, roads were not maintained, and there were no communications links with urban centres. However, for people who had lived through decades of political unrest and violence, the current relatively peaceful conditions were a great advance, and were perhaps the most significant improvement in village life for many years:

We are hoping Museveni will stay in power because he has given us peace to develop our village. (Vincent Katorogo EWR=18)

These views echo a study carried out in Masaka district in 1990, in which 34% of the heads (of all ages) of 220 randomly selected households questioned saw the return of peace as the most significant improvement to their well-being since 1985 (Bigsten and Kayizzi-Mugerwa 1995).

Since a possible result of a failure to pay one's taxes was a spell of imprisonment, their payment could be considered a transfer of tangible assets from the taxpayer to the state, made with the express aim of avoiding imprisonment. However, the value attached to the reported improvement in security suggests that the payment of tax may be more accurately seen as an investment in one's portfolio of intangible assets. In this case one paid one's taxes in return for the experience of living in a relatively untroubled environment. Once again, this is an example of an economic decision being made on other than purely economic grounds.

Summary

In the absence of material support from the state or non-government organisations, the primary sources of support for elders in Kikole lie within their community. The community does not possess the level of organisation to provide institutional support to its members, who therefore develop their own support networks with fellow community members. Networks are constructed with both friends and kin, the relative importance of each varying throughout the lifecourse. In old age, the supportive significance of friendships declines, and adult children become the primary source of support.

When an adult child is prepared to support a parent, the extent of support given depends on that child's circumstances. In turn, these are largely a product of the rural economic environment. It is difficult for young people to make a living in Kikole today, and this inevitably limits the support they can offer their parents, who are themselves subject to the same economic constraints when trying to maintain their own well-being. Rural-to-urban migration is a common response of young people to their economic difficulties, an action which, since their support comes primarily from children who live nearby, compromises the well-being of their parents. Those without children living nearby, and therefore least able to obtain support when in need, are more likely to be female, and to be not married.

In order to develop an understanding of prevailing cultural and social norms, the above discussion has taken no account of the current HIV/AIDS epidemic in Uganda. However, this epidemic is a significant event in the lives of most elders in Kikole, having caused the illness and deaths of many children and the disruption of their highly evolved and complex social support mechanisms. The multiple impacts of HIV/AIDS are explored in detail in the following chapter.

Chapter 7

The Impact of the HIV/AIDS Epidemic

Bantu abaafa! People are dying! (Florence Mugasha EWR=11)

Emiti emito e jigumiiza ekivira. Young trees make a strong forest.
(*Kiganda* proverb)

Transmission of HIV occurs during the transfer of infected body fluids from an infected to an uninfected individual. It occurs between partners during sexual contact, from mother to child during the perinatal period, during blood transfusion or through the use of contaminated medical equipment, particularly needles and syringes. According to the WHO's classification of HIV transmission (Mann *et al.* 1988), transmission in sub-Saharan Africa follows pattern 2, in that it occurs primarily during heterosexual intercourse. This contrasts with pattern 1, seen mostly in developed countries, in which transmission takes place predominantly during homosexual contact between men (Chin 1991). Pattern 1 transmission results in the majority of infections being seen in homosexual men, while Pattern 2 leads to approximately equal numbers of infections in men and women, as seen in sub-Saharan Africa. Significant perinatal transmission also occurs (Kengaya-Kayondo *et al.* 1995, Mulder *et al.* 1996), but homosexual contact has not been documented to any great degree among persons with HIV seropositivity in Africa (Chin 1991). Blood transfusion is the third most significant mode of transmission in Africa (Jager, Jersild and Emmanuel 1991). Injecting drug use is not a major problem in sub-Saharan Africa, but in view of equipment shortages and the presence of large numbers of unqualified biomedical workers, transmission through the use of contaminated syringes cannot be dismissed (Piot, Goeman and Laga 1994).

The progress of the HIV/AIDS epidemic in Uganda

The first confirmed cases of AIDS in Africans were reported in 1983 (Clumeck *et al.* 1983, Offenstadt *et al.* 1983). These were African men living in Europe, having travelled there from Rwanda and Zaire. Investigations in these and other African countries revealed the presence of people with AIDS; people with the symptoms of AIDS had been seen in southwest Uganda in 1982, and in Masaka in 1983 (Bond and Vincent 1991). Serwadda *et al.* (1985), reported that the disease was known locally as "slim disease", since the major symptoms were weight loss and diarrhoea. At that time, the origin and modes of transmission of the virus were the subject of speculation, as were both the possibility of the development of a cure or a vaccine (Biggar 1986), and the future course of the epidemic (Nabarro and McConnell 1989).

Into this vacuum came sensationalist predictions of the scale and impact of the epidemic (Hooper 1987), followed by accusations that these predictions were built on scientists' racially prejudiced and ethnocentric preconceptions of African people and their culture (Waite 1988, Chirimuuta and Chirimuuta 1989). These accusations were countered by researchers who were concerned about the potential scale and impact of the epidemic if African governments did not face up to the implications of the presence of the virus in their populations (Carswell 1988).

In 1987, Uganda became one of the first countries in Africa to collaborate with the WHO in setting up a National AIDS Control Programme (Goodgame 1990). This program has facilitated the monitoring of the progress of the epidemic, and the development and implementation of preventive interventions. The major route of HIV transmission in Uganda has been found to be heterosexual contact (Sewankambo *et al*. 1987, Malamba *et al*. 1994). Since few people in Uganda obtain an HIV test, the numbers of reported cases of AIDS grossly underestimate the extent of the epidemic. Those who have not been shown to be HIV positive, but who attend a health facility with an AIDS related illness may be diagnosed as suffering from that illness rather than HIV infection or AIDS. Larger numbers are likely, through poverty, to fail to obtain health care, and, therefore, not come to the notice of the health services. Of note here is that HIV/AIDS is not among the eleven conditions most frequently reported by outpatient attenders, although it is listed in the same document (Ministry of Health 1991, cited in World Bank 1993b) as the second most serious cause of mortality in Uganda, being responsible for 9.3% of deaths occurring in the nation's hospitals.

At the end of 1997 computer modelling of the progress of the epidemic in Uganda estimated 430,000 women and 440,000 men aged 15-49 years to be infected with HIV (UNAIDS 1998), figures which are equivalent to a prevalence of HIV infection in this age group, of 9.5%. At this time, some 67,000 children aged 0-15 years were said to be HIV positive. The probability of a child becoming infected through parenteral, or non-sexual casual or household contact, or through insect bites, is remote, and infection in children is likely to be due almost entirely to mother-to-child transmission (Mulder *et al*. 1996). UNAIDS also estimates that 1.7 million Ugandan children under 15 have lost their mother, or both parents, since the beginning of the epidemic, of whom 1.1 million were alive at the end of 1997 (UNAIDS 1998). In 1996, the average life expectancy in Uganda was about 40 years, approximately 13 years less than it would have been had the AIDS epidemic not occurred. This discrepancy is expected to have increased to approximately 20 years by 2010, by which time life expectancy is predicted to have dropped to about 35 years (Cohen 1998).

The epidemiology of HIV and AIDS in the study population

Kikole is located approximately five kilometres from Kyamulibwa, the fieldsite of the Medical Research Council (UK) Programme in AIDS in Uganda. This programme has, through demographic, medical and serological surveys, studied the progress of the HIV/AIDS epidemic in its cohort of some 10,000 individuals. Since its inception in 1989, the programme has produced some of the most significant African data on this subject. As the two fieldsites and study populations are very similar, I am confident that data concerning the prevalence and incidence of HIV, and the risk factors for becoming HIV positive in the MRC cohort, can be considered applicable to the population of Kikole.

Within the MRC cohort the prevalence and incidence of HIV are approximately 8% and 1% respectively (Mulder *et al.* 1994), and half of all deaths of those aged 13-44 years are HIV-related (Mulder *et al.* 1994, Kamali *et al.* 1996). MRC researchers have published many papers concerning the risk factors for becoming HIV positive. Nunn *et al.* (1994) found that those most at risk were young married women who, the researchers presumed, had commenced sexual activity recently. Others with relatively high risks were those who travelled away from the village, and those with genital ulcerative disease. Uncircumcised male status has been suggested, by others, as a risk factor (Bongarts *et al.* 1989, Caldwell and Caldwell 1994), and Muslims, possibly because of their higher incidence of male circumcision, are at a lower risk of infection than non-Muslims in the MRC cohort, who do not practice male circumcision (Nunn *et al.* 1994, Kengaya-Kayondo *et al.* 1996). Those with many partners are at greater risk (Malamba *et al.* 1994). Kengeya-Kayondo *et al.* (1996) found the Baganda in the MRC cohort to be at lower risk than the non-Baganda, and hypothesise that this is a result of their relative wealth. Seeley *et al.* (1994), exploring further the relationship between HIV infection and poverty, showed that the male and female heads of the poorest households, with poverty being assessed through a household item index obtained during a community survey, had a higher rate of HIV seropositivity than other heads of households.

Incidence and prevalence of HIV infection among members of the MRC cohort is less than for younger adults, as shown in Tables 7.1 and 7.2. Prevalence rates have not shown a significant decline between 1990 and 1997. This is likely to be a reflection of the long-term nature of HIV related illness, although survival time after infection with HIV in Africa has been little researched. A study of the natural history of HIV infection among a group of HIV positive members of the MRC cohort has produced an estimated median time for progression to AIDS by prevalent cases, i.e. those found to be positive at the start of the MRC study, of 4.3 years, and a median survival time after progression to AIDS of 9.3 months (Morgan *et al.* 1997). These are shorter periods than those seen in developed countries, where drugs are available to treat opportunistic infections among those who are HIV positive and who have AIDS. These drugs are generally not available in Africa (Morgan *et al.* 2002). However, the participants in Morgan's study were treated for medical conditions found during the study, and this treatment may have resulted in longer survival times than would otherwise have occurred (Morgan *et al.* 1997). Also among the MRC cohort, Nunn *et al.* (1997) report life expectancy from birth in the total population to be 42.5 years, and 58.3 years in those known not to have HIV infection.

Table 7.1   Overall incidence rates of HIV among males and females in the MRC cohort, 1990-1997

Sex	Age in years	Number of cases	Person years of Observation	Incidence rate, per 1000 person years	Confidence interval
Male	13-24	42	7219.8	5.817	4.299 - 7.872
	35-59	25	2797.3	8.937	6.039 - 13.226
	60+	7	1504.8	4.652	2.218 - 9.758
	Total	74	11521.9	6.423	5.114 - 8.066
Female	13-24	65	7228.4	8.992	7.052 - 11.467
	35-59	13	3430.1	3.790	2.201 - 6.527
	60+	4	1398.0	2.861	1.074 - 7.623
	Total	82	12056.5	6.801	5.478 - 8.445

Source: Carpenter 1998

Table 7.2   HIV Prevalence among the MRC cohort, by age and sex, 1997

Age	Prevalence rate, %		
	Men	Women	All adults
13-34	5.19	8.82	7.03
35-59	11.26	6.92	8.88
60+	1.93	1.01	1.48
Total	6.31	7.51	6.92

Source: Carpenter 1998

As discussed above, many young people leave the village looking for work, and the initial destination of most is Kiguma. Here, sex-work is a major source of income for women (Pickering *et al.* 1997), and HIV prevalence rates are high, at 40.4% among all adults, peaking at over 50% among women aged 20-34 years and among men aged 35-44 years (Nunn *et al.* 1996).

AIDS-related knowledge, attitudes and practice

Those aged 50 and over have been omitted from the UNAIDS/WHO estimates of adult infections on the grounds that "the vast majority of those who engage in substantial risk behaviours are likely to be infected by this age" (UNAIDS 1998 p3). This is probably true, but their omission from this analysis can only serve to reduce their presence in the minds of those who might otherwise provide supportive interventions in response to the difficulties faced by elders as a result of HIV and AIDS. These difficulties, when experienced by the aged in developed countries, have received considerable attention (eg Crystal 1989, Linsk 1994, Kaufmann 1995) but the problems of elders facing the impact of HIV/AIDS in developing countries are less well documented. The stereotypical belief of the sexually inactive older person has obstructed the development of HIV- and AIDS-related health information directed at the aged in developed countries (Kaufmann 1993). This belief has also been noted in Botswana (Tlou 1996), and persists in some recent Ugandan research. Ntozi (1997a) suggests that the failure of a survey to identify individuals who were both HIV positive and aged over 55 years could be accounted for by the fact that "AIDS patients who contracted it when they were sexually active would have died by age 55" (p7).

The fact that there is a measurable incidence and prevalence of HIV infection among the aged in the MRC cohort, in which infection is almost invariably through heterosexual contact, challenges this stereotype. Although the sexual lives of elders have been little explored, one study has found that, in Botswana, 54% of men and 30% of women aged over 60 were sexually active (Tlou 1996), and this lends weight to the assertion that older people are at risk of HIV infection and therefore in need of appropriate information to protect themselves.

The nearby and long-term presence of the MRC Programme has presumably raised awareness of AIDS issues in Kikole. However, the aged, through their lower mobility and social activity, may not have been influenced to the same extent as younger people. The government health services have held seminars in the area with the aim of informing people about HIV and AIDS, but one participant thought that old people were reluctant to attend these:

> We old people don't think we are at risk, so we don't go to the seminars. And those that have casual partners are scared to ask their husband or wife to use a condom because they would say: "Do you think I have the infection, or are you having other partners? You'd better leave me!" People wouldn't want others to know if they went to the seminars. (Ibrahim Sabiiti EWR=8)

Other comments suggested that the seminars had led to dissemination of HIV-related information through the village. For example, both an unmarried woman and a married man had decided on a strategy to protect themselves from HIV infection through heterosexual contact:

> I didn't go to the seminar, but I heard about it from people who did go. I decided to stop having sex − I can't be killed by it if I don't let it into my body.

We hear that the best way to protect yourself is to stay with one partner, as we have done. If you play sex with many people you get infected. When young people get married the priest tells them to stay with one partner as God has instructed.

Another man had learned of the dangers of unsterilised medical equipment, and was concerned about the risk of attending local health facilities:

You should always go to the same health worker, and you should take your own needle with you, and boil it when you get home. At big hospitals they know about this, about sterilising.

Condoms were thought to be inappropriate for use by old people, who rejected them in favour of monogamy:

I don't know much about condoms, and I've never seen or used one. I think it would be risky to use; it might get broken and I could leave a piece of it in the woman. I think it would be better for me and my wife to remain faithful to each other.

The MRC Programme, operating from a nearby village, provides free HIV testing on demand to the residents of Kikole. Few participants had taken advantage of this offer, and negative results were treated with suspicion:

I have had two tests and they were both negative. But we think they don't tell us the truth because if they did we would worry and die sooner.

Information on the current HIV serostatus of elders in Kikole was unavailable. A few of them had had an HIV test, but none in the recent past. Whilst some said they had stopped all sexual activity, a few revealed that they were still sexually active, and had multiple partners. The large number of deaths said to be due to AIDS led one woman to assume she was positive, even though she had not been tested.

We old people have it as well. In fact I suspect I am positive because one of my partners has died. I won't get a test, because I know in my heart I am positive. Don't be surprised when you hear that I have died!

The fatalistic acceptance of this woman was shared by a male participant:

These days I don't think anyone is negative any more, except those who have never been to a clinic or who don't have sex. The rest must be positive! ...I think I must be positive, like everyone else.

Younger adults were of the opinion that old people, both men and women, continued to be sexually active and to be at risk of AIDS:

A person can be old, but still want to have sex, and they say to a young girl "Ha! If I wasn't old you'd take more notice of me!". But when an old man has some money he can say to that girl "I want you, come and sit next to me." Then they will talk and he will tell her that he has Sh10,000 with him. So, an old man could sell his coffee and then try and attract a girl he wants by giving her all the money from the whole season's coffee! That woman wouldn't mind then that he was old, but would go with him because of his money, and if she has AIDS she will infect him. He might abstain again after committing this crime, but if the girl was infected then he gets the infection anyway. A man may look old, but be cunning, still behave like a youth!.... (Young man)

An old man who lives alone may decide to have sex with an old woman, thinking he'd be safe with her, but old women have sex with young men these days, and she could end up infecting him. (Young man)

It is of interest here that, in both these scenarios, the woman is seen as a source of infection for the man, rather than vice versa. In Kikole this is a widely held opinion, among both men and women, although research indicates that, in rural areas, most new infections take place from male to female (Serwadda *et al.* 1995).

Community attitudes to people with AIDS

After transmission, and over a period of years, the virus damages the body's immune system and renders the individual vulnerable to opportunistic infection by organisms to which he or she is exposed. In Uganda these most frequently and visibly manifest as the weight loss and diarrhoea noted by Serwadda *et al.* (1985). In view of the high incidence of AIDS in the area, its symptoms are well known to local people. Loss of weight, the presence of rashes or sores, changes in skin colour or in the texture of the hair were all believed to be signs of AIDS, and made it impossible to hide one's illness.

Stigmatisation has been identified as a problem for people with AIDS: as disapproval for having engaged in unsafe or forbidden sexual relations, as a reaction to the often highly visible signs of opportunistic infection, or through the labelling of people with AIDS as members of a "high risk group" (Goldin 1994). Muyinda *et al.* (1997) explored the stigmatisation of people with AIDS among the MRC cohort. They note that, in the early stages of the epidemic, it was said to have been shameful to have AIDS, but that, today, a more tolerant attitude is apparent. They ascribe this change to the provision of counselling services and home care advice. A similar change was said to have occurred in Kikole:

People are sad, but not ashamed. They don't feel guilty because they are not the first person to have it. It's so common to have AIDS that it's natural! In the beginning of the epidemic people thought it was contagious and were afraid of catching it. People who were ill were left alone with the door closed, but this doesn't happen any more. (27 year old man)

Although this man is of the opinion that familiarity with the disease has reduced stigmatisation, government seminars and the availability of testing and counselling through the nearby MRC programme are also likely to have contributed to this behavioural change. In spite of this reported change in attitude, some people were reluctant to reveal the true cause of death at a funeral:

> At a burial they say that the person died of what actually killed them, say malaria or diarrhoea. But everyone would know that they died of AIDS. Everyone knows who is positive and in the end they die of AIDS. If a person plays sex with another, and one of them dies then everyone assumes the surviving one has AIDS. (35 year old man)

I have chosen to protect individuals' confidentiality by not attaching identifiers to most quotations in this chapter.

The impact of the epidemic on the participants' households and families

Table 7.1 has shown that, among the MRC cohort, very few adults aged 60 years and over were found to have HIV infection. During this study none of the participants suffered, or died from, an illness that was said to be HIV-related. The following discussion of the impact of the epidemic on older people's lives is therefore centred on the changes they experience as a result of the presence of the virus in others, primarily their children.

Barnett and Blaikie (1992), in an analysis of the impact of AIDS in Rakai district of southwest Uganda, categorise households as having been either "afflicted" or "affected" by the epidemic. Afflicted households are those in which a member of the household has either died from AIDS, or is currently ill, while an affected household has lost support from family members who had died, or gained new dependants, such as orphaned children, as a result of the death or illness of a family, but not necessarily a household member. Barnett and Blaikie also describe a group of "unaffected" households, in which no-one was ill, or had died, and which had not been affected by the illness or death of a member of a related household.

Table 7.3 presents the impact of the epidemic in Kikole, using these categories. It shows that four elders' households were afflicted by AIDS (i.e. had lost a member of the household, or cared for a dying child). Children had not necessarily been cared for at home; one mother moved to Kampala to care for her son in hospital, while the fathers of some of the others had paid for health care in local hospitals. Eight female- and four male-headed households had been affected by the epidemic, either through the loss of a non-resident, adult child or by caring for orphans. Only one household was afflicted but not affected; here an unmarried son had died without having produced any children.

Table 7.3 shows that those households that are changed by the AIDS epidemic are wealthier than average, but except for those few households that are both affected and afflicted, the difference is small. These figures therefore, are not consistent with the suggestion that AIDS is a disease of poverty, although the special circumstances of the aged may make their experience different from that of the rest of the population.

Table 7.3   The relationship between the impact of the AIDS epidemic on households headed by adults aged 60+ and the sex and emic wealth ranking of the head of household

Type of impact	Male-headed households	Female-headed households	All households	Mean emic wealth ranking
Affected only	2	7	9	14.1
Afflicted only	1	0	1	12.0
Both affected and afflicted	2	1	3	8.7
Unaffected	8	5	13	16.7
Total	13	13	26	14.7

Table 7.4 indicates that households headed by Baganda were significantly more likely to be either affected or afflicted by the epidemic:

Table 7.4   The impact of the epidemic on households headed by Baganda and non-Baganda adults aged 60+

Impact of the epidemic	Baganda households	Non-Baganda households	All households
Either affected or afflicted*	8	5	13
Neither affected nor afflicted	3	10	13
	11	15	26

* Chi-square = 3.94, p<0.05 (uncorrected)

This result also runs contrary to the assertion that AIDS is a disease of poverty since, as shown in Table 3.3, the emic wealth-ranking exercise found the non-Baganda participants to be significantly poorer than the Baganda, a disparity which can be plausibly extended to their children. Several explanatory hypotheses present themselves: the first is that the Baganda participants have more children than the non-Baganda, as shown in Tables 6.3 and 6.4. If all children experience a similar risk of infection the Baganda are, therefore, more likely to lose a child. Secondly,

the Baganda participants have had more relationships that produce children than
have the non-Baganda (see Table 2.14). If this applies equally to their children,
then the Baganda may be "at risk", through having more sexual partners than the
non-Baganda. Thirdly, some of these deaths occurred in the early years of the
epidemic, at a time when participants said that the wealthier members of the
community were most affected by AIDS. Finally, it is possible that poverty reduces
the frequency of travel to urban areas, where the prevalence of HIV is higher and
the risk of HIV transmission greater. The last of these, if correct, implies that
whilst AIDS is, indeed, a disease of poverty, a certain level of wealth is needed
before one's risk becomes significant, and so the very poor may be less at risk than
the merely poor.

Female-headed households are more likely to be either afflicted or affected than
those headed by men, as shown in Table 7.5:

**Table 7.5   The impact of the epidemic on households headed by men and
women aged 60+**

Impact of the epidemic	Male-headed households	Female-headed households	All households
Either affected or afflicted	5	8	13
Neither affected nor afflicted	8	5	13
Total	13	13	26

Since it is normal for a woman to be younger than her husband, she is more
likely than he to still be alive when her children are old enough to become sick and
die from AIDS. Most women participants in this study produced their last children
at least twenty years ago, and all of these can now be considered to be a risk of
AIDS, whereas the aged men are still producing children with younger wives, and
many of these are not yet old enough to be considered to be at risk of death from
AIDS. Five of the women heads of afflicted or affected households are widows and
their experience fits this pattern. The other three are younger women living apart
from the fathers of the children who have died. Two of them are women who have
chosen to be unmarried, and the third was expelled from her husband's home. Since
none of these fathers was living in, or near, Kikole it was not possible to learn
whether they were also facing hardship as a result of their children's deaths, but the
experience of old men in Kikole suggests that they are probably living with a
young wife and are thus less in need of the support their deceased children might
have provided.

As an indicator of the participants' loss of social support, categorisation as
"affected" or "afflicted" is of limited use. Afflicted households, having lost a child,

may have lost that child's support, but only in the unlikely event that he or she been providing support while still alive. In most cases, those children who had died had left their parents' household before they became ill, and become sick while living in an urban area. They returned home only when ill, or after their deaths. Further, as I describe later in this chapter, those few afflicted households that were able to purchase health care for their children with AIDS did not always do so. Similarly, affected households had not lost the support of family members, since this was not usually available, but had, in most cases, gained dependent AIDS orphans following the death of a family member. These households were then faced with the additional costs of the care of these children. The impact on the lives of old people is more accurately demonstrated by analysis of the changes in both family and household circumstances that follow the death, of AIDS, of a family member, as indicated in Table 7.6.

Table 7.6   The impact of AIDS on family and household composition among households headed by men and women aged 60+

Change in household or family composition as a result of an AIDS death	Male-headed households	Female-headed households	All households
Loss[a] of one or more sons	3	6	9
Loss[a] of one or more daughters	1	4	5
Loss[a] of one or more grandchildren	1	7	8
Supporting or caring for orphans	2	4	6
No change[b]	8	3	11
Totals[c]	13	13	26

[a] The loss here is not necessarily of a co-resident, but of a family member.
[b] The total of unchanged male-headed households is the same as that arrived at using Barnett and Blaikie's categories, but there are two fewer unchanged female headed households since two of them had lost only grandchildren, and were therefore classified as unaffected by Barnett and Blaikie.
[c] Totals do not tally as some households were affected in more than one way.

Elders are affected by the epidemic more through the fulfilment of their parental obligations, than the loss of their children's support. These obligations involve, firstly, attempting to prevent their children becoming sick, then caring for them when they are sick and conducting of appropriate rituals after their deaths. Finally, they are required to care for their orphaned grandchildren, often for many years. The latter experiences leave the parents with a burden of sadness which can be seen to pervade the lives of many of the study participants.

The role of parents in HIV prevention

Villagers are well aware that, since HIV prevalence is higher in urban than rural areas, residence in town involves an increased risk of acquiring HIV, and that those remaining in the village may be infected by people returning from the town:

> When a young man goes to town he plays sex with a woman he doesn't know well, and then he comes back and plays sex in the village and can infect a lot of people. All from one man who went to town! The infection comes mostly from town. Boys come back looking smart and the girls are attracted to them and play sex with them. (35 year old son of Julius Kwesige EWR=20)

Elders, through their role as advisers to younger generations, have been identified as useful resources of promotion of behaviour change and thus in the prevention of HIV transmission in developing countries (Tlou 1996, HelpAge International 1997). In Kikole, parents attempted to ensure that their children were aware of the risks they faced. One strategy adopted to try to prevent a son becoming infected was to find him a wife, in the belief that this would reduce the number of his sexual partners:

> Some people advise their sons: "My son, stop thinking about the times which were good for you, and see what is happening in the world. You should look for a woman and go for a blood test. If you are negative I will help you to marry her, and then you can stop going with this one and that one, which will lead to your death while I am still loving you. After marriage try as much as possible not to have casual partners." (29 year old man)

They were not confident that their advice would be accepted. Young people exhibited a reluctance to accept the advice of elders, as has been discussed in Chapter 6:

> We talk to our children about AIDS, but they say it is just for young people to talk about. (Gilbert Kasibante EWR=5)

Younger adults expressed a fatalistic acceptance of the inevitability of their children's infection with HIV:

> When you tell your son or daughter "My child, you should behave like this.....", they reply *"Bali nsalami embawo?"* ["Am I going to live long enough for them to make timber from my body?"]. You know that what he is going to do will cause his death, but he is sure he is going to die anyway, so he doesn't listen to you. If people know they are already infected, this proverb means that it doesn't matter whether they stop having sex or not as they are going to die anyway. (40 year old daughter of Noah Sserwadda EWR=12)

The young people who did accept advice found it difficult to follow:

It's like having a stomach and a mouth – you can't abstain from food when you are hungry. If you like a girl, you talk with her and ask for sex, not knowing that she is sick. You can use condoms for one or two weeks, but as love grows stronger and stronger you decide not to use them, and if that means you will die, then so be it. Old people tell us, but the pleasures are not avoidable. (28 year old man)

Caregiving for people with AIDS

When a child in town becomes ill with AIDS, the knowledge that they will require care as the disease progresses influences the choice of where to spend their last days:

He might go back if he's not very ill, but not if he is seriously ill. No-one could help him there, but there might be friends here who could help.... (22 year old man in Kiguma)

They live in town with no-one to care for them and when they get sick they come back home. They spend all their money just getting back, and the parents have to pay for their treatment. (31 year old man)

In most sub-Saharan African countries, few of those who are sick or dying with AIDS-related illness can be cared for as inpatients, since, even when they or their families have the resources to purchase this care, the health services would be overwhelmed by their number (World Health Organisation 1991, Cabral 1993). Therefore, some countries are encouraging home care as a cost effective mode of caring for people with AIDS (Carswell 1988, Anderson and Kaleeba 1994). In Uganda, Tembo *et al.* (1994) report that over half of the medical admissions to a Kampala hospital in late 1992 were HIV-positive, while Taylor *et al.* (1996) state that, in spite of the World Bank's contrary opinion (World Bank 1993b), the health service infrastructure in Uganda has not yet recovered from the damage inflicted on it during the political instability of recent decades, and is thus unable to respond adequately to the needs created by the AIDS epidemic. That the use of home care rather than inpatient care in developing countries may, in fact, have benefits for the patient, family and the health services, is suggested by Danziger (1994), who reports that home care is cheaper for families and the health services. Anderson and Kaleeba (1994) agree, asserting that the quality of care received at home is higher than that offered to hospital inpatients.

The choice of caregiver For most Ugandans, who cannot afford to utilise health services, and for whom inpatient care is therefore not an option, the extended family is the only source of this care (McGrath *et al.* 1993). In most of the world, caring for the sick is a woman's task (Lado 1992), and the situation in Kikole is no different:

Although I support my mother with food and so on I can't look after her when she is ill. I would call one of my sisters to come, but if I had no sisters I would ask one of her friends. It could be her husband, but usually they aren't good at it. It would be a problem for a man to have to wash sheets after she had done everything in them, but a woman is different because they are used to it, it is their role to wash. (Young man)

Parents do look after their children but they complain, particularly fathers, who say "I told you not to have sex.", but mothers are kind to their children and look after them better. (31 year old man)

Among a small sub-group of the MRC cohort Seeley *et al.* (1993) found that, of the extended family members, mothers are the most frequent caregivers for their children with AIDS, with the patient's siblings being next most frequently in this position. Whilst Seeley found only sisters to be caregivers, Ntozi (1997a) found that care by siblings happens most often between brothers, who are more likely to be living near their parents' home than are sisters, who tend to live elsewhere, in their husbands' homes. Box 7.1 contains an illustration of the intergenerational and gender issues that arose during the home care of a young man with AIDS.

Box 7.1 Choice of caregiver

Boys find it difficult to be open with their parents. I have a friend who was ill with serious diarrhoea. His mother was caring for him at home, and although his parents bought tablets from the shops but there was no improvement, and the boy was refusing to eat. I went to visit him when his mother was out, and he told me that he thought he was going to die because he wasn't eating. At first he said that this was because he had no appetite, but when I asked him how he liked being cared for by his mother he said he wanted to eat but was afraid of soiling the bed which would be a problem for his mother, since he was unable to walk to the toilet himself. So I told the mother about this and she called a brother to come and care for him. Then he started to improve because he could ask his brother for anything, much more easily than he could with his mother. Now he can even walk a little again, and will probably live for some months. (25 year old man)

Ntozi's (1997a) study also reveals the lack of outside support for families providing care for their members who are sick with AIDS. He reports a 1992 survey of 77 households in Masaka district which found that 1.3% accessed health services, 11.7% used the services of The AIDS Support Organisation (TASO), 1.3% received help from friends and relatives, and 85.7% received no help at all. TASO services were not available in Kikole. The MRC employed a community nurse to provide advice to carers of those in the late stages of AIDS, but she did not attend any of the participants during the study. Aside from this MRC input, it can safely be assumed that households in Kikole received almost no support when caring for people with AIDS. Seeley *et al.* (1993) go further, and assert that, within a household, the burden of care often falls upon an individual who will not receive support in this task from other household members. Ntozi (1997a) notes that daughters find caring for fathers very difficult, and will not do so. In these situations the burden of care is likely to be transferred to their aged parents, an event which illustrates well the role of elders as a resource to be exploited when others have been consumed or are, for any reason, unavailable. During the study it was not unusual to hear a participant, when presented with a problem by a son or daughter exclaim: "What can we do.....what choice do we have?"

The provision of care   Early in the epidemic, knowledge about AIDS had been limited, and this had influenced the care provided to people with the disease:

> My son had gone to live at Diimo, on the lake shore near Kiguma, looking for work, and he was already weak when he came back here. These patients need to be fed good food, but he was among the first people to have this illness and we didn't know what was the right thing to do for him. We took him for injections and tablets and *kiganda* medicines and it was really a burden for me until he died. I spent all the money I had on medicine. Once I was away for two weeks at a traditional healer's place, looking for a cure.

Today, although the symptoms that occur during the advanced stages of AIDS are well recognised, and presumptive diagnoses are made by those familiar with the sick person's sexual history, the earlier signs of the illness may be confused with other, traditional diseases:

> At first my wife's daughter was in Kiguma, where she was married. Her husband died and she stayed on there for sometime before she became ill. After a month we realised that she had no help there, and so we moved her back here. She was ill for three months before she died and we were greatly affected by this. It is difficult to know which disease is caused by a charm, and which is AIDS. My wife said that her daughter had been charmed by the women that shared her rented house. If someone complains that she has been charmed you automatically take her to a traditional healer, and we took her to a man who has a research clinic in Masaka, where she was given traditional drugs which she took until she died. We wasted a lot of money on them. When we realised she had AIDS it was too late to take her to the MRC, or anywhere else, because we had run out of money. Because she had AIDS she needed to eat well, things like fish, eggs, sugar and salt. I did all this and now I have a number of debts to pay....It will take a long time but they know I have only a little money and I think they will be patient....

Once the illness was recognised as AIDS, this man had wanted to obtain biomedical treatment, indicating that this was seen as the most appropriate treatment for this disease. In the following instance, adult children who returned to die in their mother's home were taken directly for biomedical treatment:

> I have cared for two sons and a daughter. It was two years ago now. They lived in Kiguma, but when they became sick they came back here. They had AIDS and it was difficult to care for them because I didn't have much money.....I could get a little money from my *mbidde* plantation, but I had to buy milk, passionfruit, meat and sugar....I took them to Kyamulibwa sub-dispensary and to Vila Maria hospital. It was very expensive but I had no choice. I borrowed money from my friends who would lend me a little to pay the hospital

bill, and friends of my sons and my daughter would visit and bring a little money for passionfruit or something like that. I also had to meet the costs of the burials, and I am still paying them off now, two years later!

As these examples indicate, parents usually cared for sick children in their own home. Only one woman participant had left the village to care for a child who was dying elsewhere:

My son was living in Kampala when he became ill and he went into a hospital there. I went there and cared for him for the last month of his life. The hospital staff gave him medicines, but I did everything else for him; washing, cooking, and I had to buy his food. He stayed there until he died, which was very expensive. I went there without enough money and no-one helped me. In our family each of us struggles alone. I had to borrow the money, but I paid it all back when I sold his property after his death.

During the study two of the participants were required to care for people in the final stages of AIDS, when very sick younger relatives returned to Kikole from urban areas shortly before their deaths. One was Maria, who had separated from her husband some years earlier and had a home in Kiguma. On returning to Kikole she was cared for in her mother's home. The other, Paul, was a young man who had been living in Kampala with his wife and children. He was cared for, in Kikole, by his mother, but in the home of his grandfather. In both these situations the choice of a place to die was made, without consultation, by the sick person, following which the burden of care fell, again without consultation, on the aged relative.

Maria had one daughter in Kampala and another in Ggomba district, and a son who had no fixed residence. She had been ill in Kiguma for a long time, and when she realised she was dying she came to Kikole to be cared for by her aged mother. When she arrived at her mother's house Maria was already very ill. Her skin was discoloured, her hair very thin, she was emaciated, and barely able to stand. Although these symptoms were widely accepted as indicative of AIDS, her mother, possibly fearing stigmatisation, did not acknowledge that Maria had AIDS, instead claiming that she had malaria. During Maria's illness one daughter arrived from Kampala, with a newborn baby. She was unmarried and penniless, but helped with the physical care of her mother.

Maria's mother, one of the poorest women in this study, lives alone, but close to her very old and frail mother. Maria brought no funds with her, and her mother had very little money for health care. Visiting the sick is a community obligation, and although many people visited with good wishes, her friends and neighbours provided little material help. Maria suffered from continual vomiting, and I gave her mother money to buy an injection to help with this problem. This would only have been a temporary solution to an ongoing problem, and as such it may not have been an appropriate gesture. In the event, she purchased tablets which were cheaper, but ineffective. My later gifts were of soap and sugar, intended as much to help her mother as to ease Maria's condition:

I bought some tablets but she vomited them. I don't have the money to pay for an injection, and she is still vomiting whenever she eats anything. I went to Tadeo's home [a neighbour] to ask for help, and he gave me passionfruit and oranges which I gave to her after adding the sugar you gave me. I didn't have any of that, I gave it all to her.

Tadeo is a coffee trader, and one of the wealthiest young men in the village. He lives next door to Maria's mother, in a brick house which has the only glass windows in the village (his house is shown in Figure 5.1). It is indicative of the paucity of support offered to her, and indirectly of the level of poverty in the village, that a simple gift such as passionfruit and oranges should be given by a relatively wealthy man and be worthy of comment.

Maria spent her last days lying under the banana trees behind her mother's house, and died without receiving any further health care, two weeks after arriving in the village. She died in the company of her mother, her grandmother, her daughter and her granddaughter. Her mother's adult grandchildren were helped by neighbours and members of her church to bury Maria. Maria's grandmother also died, only a few weeks later, and the granddaughter returned, with her baby, to Kampala. At the end of the study, Maria's teenage son was occasionally staying with her mother, and she was hopeful that he would remain in Kikole and be available to support her as she grew older.

Paul, the second person to die of AIDS in Kikole during the study, was the son of Emmanuel Kabazi, a local dignitary and successful businessman. Kabazi is the son of one of the male elders included in this study and now lives about five kilometres from Kikole. His father was therefore Paul's grandfather. One of Kabazi's brothers, and thus Paul's uncle, also lives in Kikole, close to his father. This uncle is relatively wealthy and the family is, therefore, much better placed than most to provide care. Paul had grown up in his grandfather's home and although he had a wife and children in Kampala, he made it known to his grandfather that he wanted to return here to die:

> The son of my son in Kyamulibwa is seriously ill in Kampala and will be brought here next month....He lived with me from the age of six until he went to Kampala, and he decided that he wanted to come back here. He can't stay at his father's house because as he is married there is a problem of *obukko*. His wife cannot stay there.

When he arrived he was able to walk short distances and was visited at his grandfather's home by many friends. He was very thin, and had multiple sores in his scalp which caused him great discomfort. He always wore a hat, which he said served to keep insects away. He did not mention that it also served to hide his condition from others, but I am sure this was a factor since his scalp was a shocking sight. He brought no funds with him, and his grandfather was very poor, in spite of the relative wealth of his two sons, Paul's father and uncle.

> He needs medicine, and money to bring his wife and property here from Kampala. We are looking for someone to go there and collect them but we have no money.....I have no support at all from anyone, and I sent my younger grandson to ask the relatives for support, but so far they haven't come up with any. ....I am alone with no one to help me.

When the disease progressed further his mother came to care for him. Since Kikole is not situated within the MRC study area, its residents are not usually able to obtain health care at the MRC clinic, but in this case its staff did provide some outpatient care during the last week of his life. He was cared for inside his grandfather's house on a mattress covered with clean sheets and a blanket:

> My grandson is very sick. We are just waiting for the day God has prepared for him to breathe his last. He can no longer speak; he just groans now, and I don't think he will live much longer. We took him to the MRC for treatment a few times. We used to tie a chair on the carrier of a bicycle and take him down there, but he is too sick to do that now.

The sick man's wife visited, with their children, from Kampala, but did not stay long. She left without the children, who were from then on cared for by members of her husband's family. Paul's mother, an old woman, was no longer living with his father, who now had another wife in his house. The presence of this second wife made it impossible for Paul's mother to care for him in his father's home.

Although no more health care was purchased, the costs of caring were met by his family and neighbours:

> His father and my other son are paying for the costs. They are giving me food and sugar as sometimes we are short of food here. The villagers brought some food and paid for a month of milk for him.

The young man died just over two weeks after arriving in the village. His wife returned for her husband's burial, but did not take any of her children with her when she went back to Kampala. The wide dispersion of this family is typical of the fate of the Kikole families who lose one parent to AIDS:

> He left three orphans. Two are in school and live with my daughter now. She is a nun in Vila Maria. The youngest one is only a year old and is living with its grandmother, the mother's mother. The wife has gone to work as a housegirl near Kampala.

Davachi *et al.* (1988) reported that in Zaire, in the early stages of the epidemic, the average cost of the hospitalisation of a child with AIDS was three times the average monthly income, and therefore it is not a surprise to find that poor people in Kikole are unable to pay for care for their relatives. However, although neither of the two people whose experiences are described above received more than minimal health care, in the latter case the family was wealthy, confirming Ankrah's (1991) proposal that, in the last stages of AIDS the inevitability of death renders major expenditure unnecessary or unwise. Paul's grandfather acknowledged caregivers' helplessness when he advised Maria's mother that there was nothing to be done for her daughter:

> Her mother asked me to advise her on what to do for her daughter. But she is dying and there is nothing that can be done because there is no cure for this epidemic.

Household food security

The impact of the epidemic on food production in Uganda has been well documented (Barnett and Blaikie 1989, 1992, PANOS 1992, Hunter *et al.* 1993, Topouzis and Hemrich 1993, Haslwimmer 1994, Barnett *et al.* 1995). Agriculture in Uganda is labour intensive, and the loss of the labour of household members with AIDS, whether during the course of their illness or after their deaths, reduces household ability to grow food and to transport it to market. For many households, poverty prevents the employment of labour or the use of other agricultural inputs to maintain production at this time. As Hunter *et al.* (1993) note, and as I have described above, declining capability in old age reduces the ability of an individual to labour and to maintain his or her household food security. Elders, therefore, become increasingly dependent on adult children to provide food, and any reduction in these children's ability to do so will reduce the support that they can provide their parents. Thus, by compromising food security, the AIDS epidemic is exacerbating the difficulties that I have described in previous chapters, rather than creating a new problem for elders in Kikole.

Whilst those who were physically able to cultivate their land and grow enough food to feed their households did not experience a reduction in their food supply when their children's labour was lost, those who were dependent on these children, or who had been expecting to become dependent on them as their own abilities declined, were distressed by this loss:

> Our sons could help dig our *kibanjas*, but instead of working for their fathers they go to town and come back sick or almost dead with AIDS!

> The people that would have supported me are all dead! My sons have died more than daughters. So far it is four boys and two girls, and their children too!....If they hadn't died I would still be able to buy myself meat and fish and milk.

> I am afraid for my son and I pray to God to protect him. He has built a house next door and brings women home at night, and they are all full of Slim because all the young people have it. If he gets it who will care for me? I can't get water and firewood for myself, and my wife is handicapped, so if he becomes ill I will have no choice but to commit suicide!

Household access to markets

The first people in Kikole to become sick with AIDS were the wealthier and mobile businessmen of the village:

> In the early part of the epidemic rich people were dying more than poor because they were mobile. They were the ones moving from the village on business, say to Rakai, where there were people dying. So those people got it first and brought it to Masaka where the prostitutes got it, and then it came to the village, where poor people got it as well, and now everyone has it. (31 year old man)

The loss of these people had a dramatic affect on the village, which has been deprived of those individuals with the ability to initiate development and thus improve the lives of its residents:

> The village was developing, but so many people are dying that it has stopped. They are all dying, young, middle aged and old. In the past there wasn't this AIDS epidemic.... It's the rich ones who could develop the village who are dying.

More specifically, their loss has reduced the ability of farmers to access the market:

> These people had cars, and if you needed to you could rush and hire one. You can't do that any more because after they had died their wives moved to Kiguma or Kampala and took the cars with them. The cars used to take our produce to market and we could get cash but now we can't do that. There isn't a single motor vehicle here any more.

The inequitable access to markets of village people, described in Chapter 4, is largely a result of this lack of transportation. They are now in the disadvantaged situation of being dependent on buyers to come to them, and of being forced to accept the price offered by them, rather than of taking their produce to market and selling it to the highest bidder.

As reported in the previous chapter, young people leave the village because of its lack of economic opportunities, and this lack must be worsened by the absence of its entrepreneurs, who have died from AIDS. Thus, since support for elders comes primarily from those living nearby, the epidemic has reduced the support networks of old people by killing entrepreneurs and forcing adult children, who may not be HIV positive, to leave the village in search of work.

Expenditure of time and money on burials

Elders' land often includes a family burial ground, and clan descendants of the head of the original head of the household are customarily buried there. Thus, participants were obliged to bury not only their children, but also their grandchildren. It was not unusual for a grandparent to be responsible for the burial of a grandchild who had been born and died in Kampala, and whom he or she had never seen alive. With fertility rates remaining high, a grandparent could have several dozen grandchildren, and be liable for their burial costs in the event of their deaths. Exceptionally, a widow was also responsible for a great-grandchild's burial:

> I am the mother of the grandfather of the child we buried last week. His father and his grandfather also died recently. They all died of AIDS. We rarely go two months without a burial in this household. So far we have lost ten people to this disease.

Burials are increasing in frequency. Since kinship and friendship links extend beyond the boundaries of Kikole, individuals were expected to attend burials being held at the homes of many relatives and friends:

Only last week we had about six burials. We spent the whole week burying people! People used to die but not in the numbers they do today. Today it is mostly the young people that are dying, because of this epidemic of AIDS.

Roscoe (1911) describes *kiganda* mortuary rituals that are required after the death of a peasant. After the burial a period of mourning, with accompanying ritual, lasted about a month. Mourning formally finished at the time of the *orumbe* ceremony, during which the installation of the heir to the deceased person took place. Mair (1934) reported that these rituals were largely unchanged, but Barnett and Blaikie (1992), working in Rakai in 1990 and 1991, found that in response to the growing numbers of funerals taking place, funeral ceremonies had reduced from a length of three days, to one of only a day and a half.

In Kikole, as in Rakai, there is no expectation that mourning will continue until the installation of the heir, but it is still considered respectful to stop work for a period after a death.

The family don't work for a while, maybe a week, after a death. If this happened during the planting season some people end up not planting! Sometimes neighbours will visit in the evening to try to make the bereaved people happy, and other relatives may come and stay for quite a few days for the same reason. So they miss out on digging as well. (45 year old man)

However, this practice was seen by some as an inconvenient interruption to essential economic activity and is avoided if possible:

When someone dies during the night relatives move through the village telling everyone, and then in the morning people have to go and say their farewells and sympathise with the family. If you are living nearby you aren't allowed to do any work that day, not even pick coffee, but only to get your food. But if you live a long way away and the family won't know, then you can dig. If someone dies during the day then neighbours have to go and spend the night at his or her home. If you don't then no-one will come to your compound when you have a death and you will have to sleep with the body and dig the grave alone. (Noah Sserwadda EWR=12)

The *orumbe* was sometimes held on the same day as the funeral, saving both time and expense. This behaviour was disapproved of by those who could afford to do otherwise, in a revelation of another of the few sources of inter-tribal tension in Kikole:

Baganda are not allowed to have the *orumbe* on the same day as the burial, but this doesn't stop the others doing it! (80 year old man)

As discussed above, every household is expected to contribute towards the costs of each burial that takes place in the village, an obligation which becomes difficult to meet as the frequency of burials increases. The cumulative load of many obligatory payments may exceed the financial resources of the household:

In the past people used to notify neighbours and relatives in other villages by note – a chit was passed around from person to person. That was done free, for nothing, like everything related to a burial. No-one charged for their help. There were problems though for people who couldn't read. But these days rich people are using megaphones. These have batteries and have to be paid for – up to Sh5000 a day and Sh2000 more if you want to hire someone to go around with it as well. People have made a business out of this and it's very expensive. (31 year old man)

Poorer people, when they cannot afford all the condolence fees, stop going to all the burials that are announced. Rich people are invited to many burials, in the hope that they will provide support. One of the wealthier participants complained that attending many burials was causing him economic difficulties:

I've been to nine burials since you last visited me [a month ago]...... Going to all these burials affects me in three ways. I can't manage to keep up to date with the work on my land, then I have to pay a condolence fee at each one I go to, and finally all the travelling makes me tired so that I can't work when I get back.

A woman said that her social life was interrupted by the obligatory attendance at burials:

People don't sit and weave mats together any more. I think this is because people are dying so often these days and my friends are away at funerals more often. You can be sitting at home and hear a megaphone announcing a burial, and whatever you are doing you have to stop and go to the burial; and the same might happen the next day! I rarely go four days without attending one, and people no longer have time to sit and weave mats.

As Barnett and Blaikie (1992) note, the burial ceremony is an occasion for the acknowledgment of the deceased's contribution to his or her family and community, and it is considered desirable and prestigious to hold an elaborate and expensive funeral for one's parents:

A man in Kalungu had 60 children, and when he died people even came from Tanzania – some were diplomats! It was the biggest funeral ever seen around here and he will always be remembered. (31 year old man)

In the past such funeral expenses could be undertaken in the knowledge that one would only have to meet them twice in one's lifetime, once for one's father, and once for one's mother. Women could expect that their brothers would be responsible for these costs. Most of the elders in Kikole today, however, have already buried their parents, and are now faced with the costs of burying their children who are dying from AIDS. Widows who are faced with the need to bury a child suffer because they do not normally inherit from their husbands and therefore have few resources upon which to call.

Since the costs of a burial are shared by all community members, and these monies are largely paid to traders who are not members of the village community, an increase in the frequency of burials has a negative effect on the economic life of the community. Poorer people can contribute less, or nothing, in spite of their concern that they may have to pay for a burial themselves, and the scale of the ceremonies is therefore reduced. Juliet compared the funeral ceremonies for her father and her brother:

> My brother was alive when my father died, and we bought a cow that time. But after my brother died last year I was on my own, and we just cooked beans.

Not only are they required to meet large, unexpected expenses, but they are also aware that there is a possibility that their children will not be alive to provide them with a burial such as they would wish and feel entitled to expect. This is a distressing prospect:

> The people who should bury us are dying! We will be alone when we die.

Personal and community grief

> The truly destructive impact of an epidemic lies in the way it strikes not only at individual life but at the very metabolism of society,....The epidemic is to be measured not in terms of loss of individuals alone, but in the progressive reduction in functioning, both of communities and families as the disease spreads. Not only resulting in lost production, food, economic activity, but also in a progressive increase in the social burden on the survivors. (Kreniske 1991 p3)

Loss is not only financial. Those who have lost children naturally also feel an emotional or spiritual loss, which may effect their well-being:

> I am not OK because of thinking about the death of my son. At times I don't eat well, not because I am ill, but because of sadness about the loss of my son.

The effects of successive deaths to AIDS are described in Box 7.2. To lose one's children not only means that one will not be honoured by a fitting funeral, but that there will be no-one left alive to record one's contribution to one's clan, or one's very existence:

> We produce our children to ensure that our clans continue. They would have been caring for us in our old age, but now we die like someone who has produced no children. For example, that land there was owned by a man who had some children, but two went away and the rest died. Then he and his wife died and the house collapsed. Now the land is fallow as if he hadn't produced children. But look at George Kabongo! He produced two children, and one of them, Ndawula, is still here, and owns most of the land now. When Kabongo dies people who go by will know that this land belongs to his son. They will know he died leaving someone on the earth. It's important that people remember you, and it means the clan will continue.

Box 7.2 "Those who have many children will cry many times."

Let me give you an example – a family near me. This man died in about 1980, before AIDS came here, and he left a number of sons and daughters, about five of them were unmarried and still at home, and some others who were married and lived away from the village. One was in Kampala and another down near the Tanzanian border. They both died of AIDS, and their orphans came back to his home, where their grandmother lives. Then a son who worked for Ugandan Airways also died of AIDS. He had two wives and was a rich man and helped his mother a lot. His orphans were brought here too. The orphans are with their grandmother in my village, but their mothers are in Kampala on their own business. Then there was another son called Mayanga who got married in 1988. His wife went to work with MRC, but she died after only a year. The man is still alive, but she left three children and they are in the same compound. Now, in the same house, is another son with seven children, and he is going to die. Then, the husband of a daughter in Masaka died and now she has some signs of AIDS. She has a lot of children who will all come to live in their grandmother's house after their mother has died. So, a family that had few children after the death of their father now has a lot of grandchildren. So we say that those who have a lot of children, say twenty will have to cry twenty times, and those who only have a few, say four children, will maybe only have to cry say four times and it's finished. (28 year old man)

This grief is not purely personal. As Campbell and Rader (1995) note, the losses to AIDS within a community are cumulative, and the widespread and continuing losses of children create a shared, or communal grieving. Seeley and Kajura (1995) have described this emotional impact of large numbers of deaths and burials among members of the MRC cohort. In Kikole, I found a shared sadness, among both rich and poor:

We all feel sad because we can't talk with the people who have died. Everyone is affected, and we are all sad.

This is a terrible situation. There is no cure for AIDS, and it will take all our children. We are crying with our sorrow.

Some were sad because of the loss of support, others because of the loss of investment in their children:

We're sad because we can't expect any help from our sons as many are dying, without even having children themselves. It's as if a banana tree has fallen without bearing fruit, there is nothing left afterwards – no fruit. But if it had matured and borne fruit then you could get some benefit from it.

One man noted the impact of so many deaths on young children:

When I was young there weren't many burials like today. When our parents had gone away to sleep at the house of someone who had died we children would be afraid that the person who had died would come out of the sitting room of his house, where he was lying, and eat us! But children of today know who is ill, who is dead and who is weak. They go to burials at a young age and are used to them. They may have seen the body of an aunt, and then the father dies and they see his corpse, then the mother and a stepmother, so they are used to it, and used to dead bodies. (30 year old man)

This omnipresence of sickness and death may explain, to some extent, the pessimism expressed by young people when discussing their futures. For example, young men in Kiguma could not see the point in returning to the village if they were going to die before they could produce crops there. This attitude deprived both their ageing parents of their support, and the village of the benefits of their economic and social activity:

We have a saying: "Why should I suffer if I am going to die tomorrow?" Very few men go back to the village, because of this infection. HIV came here, and we don't think we will live very long. It takes one or two years to get the benefit from your crops, so why plant crops that may not mature before you die? (22 year old man in Kiguma)

In 1972, Turnbull published his study of the Ik of Northern Uganda, among whom he found social institutions to be collapsing under the pressure of extreme and continuing food shortages. Whilst Turnbull's research methods and the way in which he has presented his data have been criticised (Barth 1974), his findings remain alarming, and imply that lasting stresses on a community may lead to the breakdown of the structures through which it supports its weaker members. Later, in an early stage of the AIDS epidemic, Carballo and Carael (1988) discussed the possibility of community breakdown under the stresses that might result as the epidemic progressed. Although I would not assert that this is yet happening in Kikole, the comments of two participants suggest that the multiple stresses produced by the epidemic may be compromising the caring relationships between or within families and households:

My sister is always crying, but she's crying for nothing. She's only lost three children – I've lost seven!

The grandchildren are crying and I am crying and no-one cares for anyone else....

In addition to grieving for their children, parents also worried for those who were still alive:

On Christmas day we were expecting my granddaughter to come from Kampala, but she didn't arrive. She is the daughter of my son who died, and a widow. He died of this illness that kills young boys these days, and left with six children and pregnant again. They never owned a house and lived in a rented house, and I don't know how she will manage. I am worried about her. Her health isn't good and her children aren't well either. My daughter wants to go and see her to find out what is the problem but she hasn't the money for the fare.

A young woman migrant to Kiguma, a town which survived by meeting the needs of passing truck drivers, said that parents in the village worried about daughters in her position:

> Our parents worry a lot. If we don't look well when we go back then they worry about us. And they worry that we need so much money here. Where will we get the money from? (30 year old woman in Kiguma)

In recognition of these issues TASO provides a counselling service to those able to access it services (Hampton 1991, Williams and Tamale 1991, Fleischman 1995), and the MRC programme has developed a counselling service to help local individuals, families and communities cope with the epidemic (Seeley *et al.* 1991). This service was available to the residents of Kikole, but was not accessed by any elders during this study.

The adoption of orphaned grandchildren by aged adults

Early in the AIDS epidemic it became apparent that the disease was creating many additional orphans (Preble 1990), and that the need to provide for them would present difficulties for their families and communities (Hunter 1990, Kaduru *et al.* 1990, Dunn 1992.). A particular concern was that these children would not receive adequate care from an extended family whose resources had been depleted following the sickness and death of their parents (UNICEF 1990, Seeley *et al.* 1993).

In the context of the AIDS epidemic Preble (1990) describes four possible outcomes for orphaned children: adoption by relatives or non-relatives, placement in an orphanage, abandonment, or death. Devereux and Eele (1991) suggested that as the epidemic progressed, the burden it produced on extended family networks would prevent them adopting orphans, and that the other three options would become more common. Street children are now widely seen in urban Uganda, and are said to be AIDS orphans (Ntozi and Mukiza-Gapere 1995), and there are a small but growing number of institutions providing care for them (World Health Organisation and UNICEF 1994). I was not able to obtain reliable information on the fate of the elders' orphaned grandchildren who were said to be living away from the village. Many were reported, by their grandparents, to be in Kampala, with or without their parents. Contact with them or their parents was rare or non-existent, to the extent that it was common for the participants not to be able to tell me the names of their grandchildren. Without wishing to ignore the problems of AIDS orphans, which have been widely addressed elsewhere (eg Kolsrud *et al.* 1989, Muller and Abbas 1990, Foster *et al.* 1995, Sengendo and Nambi 1997, Ntozi 1997b) the following discussion will, in keeping with the objectives of this study, focus on the experience of grandparents rather than their grandchildren.

Co-residence with grandchildren

Roscoe (1911) reported that it was normal for a child, once it was old enough to leave its mother, to be taken away from her and cared for by a member of their father's clan. Further, upon the death of a man, his heir would formally adopt his children, "making no distinction between them and his own children" (p270). Roscoe did not consider orphanhood a problem, among the Baganda, at that time:

> Children belonged to the clan, and when their father or mother died, they were still under the care of some relative who took the place of the father. (p81)

Roscoe does not discuss which relatives might assume care for young children when they were removed from their mothers. Mair (1934) confirms Roscoe's observation, and identifies the father's brother as a frequent choice, since this placement was thought to strengthen clan ties. However, Mair (1934) also notes that a man could not refuse his parents' request to send them a child, a request which might be made "as much from the desire to 'have a child about the place', as for the sake of the material advantage of its help in the house" (p60). Although Richards (1966) was "impressed with the numbers of elderly men and women....living alone or with one or two children" (p71) in a village close to Kampala, she did not stratify household composition in this village by age, and so the extent of this practice at that time remains unknown.

An analysis of rural household composition in Masaka and Rakai districts in 1983/4, at the start of the AIDS epidemic, shows that 35% of the total population were living in households containing one adult and one or more grandchildren (Jaenson *et al.* 1984, cited in Bond and Vincent 1991). Without presenting any earlier data, Bond and Vincent (1991) assert that this indicates an increase in the number of grandchildren hitherto living with their grandparents. They report that Jaenson *et al.* (1984) ascribe this trend to "a more permissive attitude toward illegitimate children of their sons and daughters" (p53), suggesting that these children may not in fact be orphans, but children born to unmarried, adult children, or to married children but by a partner other than their spouse. In 1983/4, then, AIDS orphans were not yet recognised as an emerging problem.

In 1989, five years later, by which time the epidemic was well advanced, Barnett and Blaikie (1992), showed that, among a small sample of rural households in Rakai district, orphans (undefined) were much more likely to be found in households that were afflicted or affected by the epidemic, than in those which were unaffected. They do not, however, distinguish between two- and three-generation households containing orphans, and so the extent to which these orphans are being cared for by their aged grandparents, or other aged people, remains unclear.

Also in 1989, Save the Children Fund (UK) carried out an enumeration of orphans in four districts of Uganda, including Masaka district. In this survey, they defined orphans as children, under 18 years, "with one or more parents missing through death or displacement" (Dunn 1992). Their results show that, at that time, 5% of the children in the district were orphans. Of these 59% had lost their father, 18% their mother, and 22% had lost both parents. Dunn (1992) makes the point that, since AIDS is sexually transmitted, it is likely that those children who have lost one parent to the disease will also lose the other. Approximately 20% of the guardians of these orphans were aged 60 years and over, and 31% of the orphans in Masaka district were being cared for by their grandparents (Dunn 1992).

Among the MRC cohort in 1989-90, 518 children had lost one or both parents, representing 10.4% of a total of 4,975 children aged under 15 years. 6.3% had lost only their father, 2.8% only their mother, and the remainder had lost both their parents. The HIV-status of their deceased parents was unknown, but during the following three years 169 additional children became orphans, of whom 72 (42.6%) had lost a parent who was HIV-positive. Eighty-four (49.7%) had lost an HIV-negative parent, with the remaining parents' status being unknown (Kamali *et al.* 1996).

Table 7.7   Reasons for co-residence with grandchildren aged under 15 in households headed by men and women aged 60+

Reason for presence of grandchildren in the household	Number of children by gender of head of household[a]		Totals
	Male	Female	
Requested by the head of household	6 (4)	5 (3)	11 (7)
Left by a migrating adult child	10 (5)	6 (2)	16 (7)
Born outside an adult child's marriage	7 (4)	-	7 (4)
AIDS orphans	5 (2)	14 (4)	19 (6)
Non-AIDS orphans	5 (1)	-	5 (1)
Not co-residing with grandchildren	0 (3)	0 (4)	0 (7)
Totals[b]	33 (13)	25 (13)	58 (26)

[a] Number of households in brackets.
[b] Totals do not tally since some households contained grandchildren in more than one category.

In the following discussion I shall define an orphan in the same way as Kamali *et al.* (1996), that is "a child under 15 years who has lost one or both parents". The studies described in the previous paragraphs present a number of reasons why a grandchild might be found to be living with a grandparent. Firstly, there is a tradition of placing grandchildren with grandparents, secondly there is an apparently increasing readiness of grandparents to care for the illegitimate children of their own children, and finally, grandparents care for their orphaned grandchildren, who may or may not be AIDS orphans. The frequency with which these were found in Kikole is shown in Table 7.7.

The 58 children under 15 who were living with their grandparents represented 22% of all Kikole children in that age group. The largest group of unrequested grandchildren were those who had lost a parent to AIDS, while slightly fewer had been left with their grandparents by parents moving away from the village, usually to Kampala to find work. The latter group of parents has not been identified in any of the previous studies discussed above, and since they are living in areas of high HIV prevalence their children may, in time, also may become orphans. These two groups accounted for 74% of the unrequested children living with their grandparents. Smaller numbers had been born to unmarried children, or to married children but not within their marriage, or were described as non-AIDS orphans. Table 7.7 shows that female-headed households were more frequently caring for AIDS orphans, while male-headed households more often contained other grandchildren, and further develops my earlier argument, stemming from Table 7.5, that female-headed households are more frequently affected by the AIDS epidemic than other households. However, the small numbers of households involved here renders extrapolation to the wider population inadvisable.

The placement of children following parental separation or death

In Kikole, marriage is defined by co-residence:

> A wife is a woman you live with, whether or not you are married in the church, and whether or not you have children with her. There is no definite time that you have to have lived together, but it has to be more than just a casual relationship. (Young man)

Marital instability in Buganda, and the relationship between this behaviour and the transmission of HIV has been commented upon by Nabaitu *et al.* (1994), and Adeokun and Nalwadda (1997). Relevant to the current study is the pattern of placement of dependent children following the end of a relationship. Clearly, the greater the number of serial, as opposed to polygamous relationships, that an individual enters into, the more often the question of child placement following marital breakdown is addressed.

Mair (1934) writes that Baganda children, being members of their father's clan, were invariably returned to the father after marital breakdown, regardless of the cause of the end of the relationship, while Devereux and Eele (1991) report that after the death of a father the children remain with their mother, and leave with her should she migrate to urban areas or return to her natal home. I did not find either of these scenarios to be the usual practice in Kikole, where child placement decisions were made on an *ad hoc* basis, with reference to the ability and preparedness of either or both parents' family members to care for the children.

The child may continue to live with the surviving parent or be placed in another household. The norm described above (Mair 1934), according to which the child was cared for by father's clan members, was found to persist, although there are now accepted circumstances under which the mother's kin may care for the children:

> An orphan will go to the paternal uncle, paternal aunt, or the grandparents. If the mother is still alive they can be divided among her parents, but if not they will go to the father's parents. If both parents are dead then the clan members decide who will have them after the funeral rite.

This decision is made with a view to ensuring the costs of child rearing can be shared among both parents' families, if the parents' identity is known:

> If your daughter dies leaving a child with you and the father is alive, it's important to know his location because you can negotiate where the child should be brought up. He might, for example, continue to live with you, but if he dies or gets ill, then the father could help financially, or take part in discussions about what to do – whether to get treatment or where to bury it.

Paternal kin treat the children of a son differently from those of a daughter, since the former are members of their clan, and the latter are not.

> When I talk of grandchildren I mean those of my son. Those of my daughter would go to her husband's place when they get a bit older, whereas those of my son to whom I have given land, can grow up with me, their grandfather.

This elder alludes to another norm: that the widowed or separated mother of young children will continue to care for them until they are independent, after which time they may be reclaimed by their father or his family. Since she has lost the support of her husband and is not considered a member of his family, the mother usually returns to her parents' home with her young children, whereupon her parents become responsible for both her and her children:

> I am looking after my daughter's son as if I am its father, and the real father is doing nothing. The baby is just sucking from me, like sucking from a straw! It is a burden for me which I did not create for myself.

Following the death of one parent or marital breakdown, there was concern among parents and grandparents that, should the children remain with their mother or father and a step-parent, the step-parent would neglect or abuse them, discriminating against them in favour of the younger children of their own marriage. This applied both to serial marriages, where the surviving parent has remarried, and the new spouse rejects the children of the previous marriage, or, as in the case below, in polygamous marriages, where a surviving wife rejects the children of a deceased wife:

> A grandson has come to live here. He is 14 and his mother was my daughter and has died. I think he was chased away by his father's other wife, and since his mother was

born here he decided to come here. Co-wives don't like their step-children, and so everything that went wrong was blamed on him.

To avoid this event, there was a reluctance to allow children to remain with either parent, in which case their grandparents were expected to care for them:

> My daughter produced five children with her husband in Kyaggwe. But she left him long ago, and brought the children here. She has a new family now. I didn't ask for the children to come here, but in Kiganda culture you have to take them.

Here, once again, is evidence of the powerlessness of the aged in the face of social change. This is the experience of many elders, as the changing patterns of marital relationships among the young, and the AIDS epidemic, direct their grandchildren into their households.

> If a wife leaves her husband it is the duty of his mother, if she is alive, to look after his children until they are grown up. We have a saying: "An elephant has to carry its tusks." They are my tusks!

> They are both my son's children, one is two years old and the other three. They were both produced outside his marriage. When their mother had finished breastfeeding them she brought them here and left them. We just found them outside the door one morning. We can't chase them away since they are of our blood, and their father, our son, can't take them to his wife because she didn't produce them and will say "Take them back to their own mother!".

> I have had a granddaughter come and live with me. She has come from Kampala with two of her children. She has separated from her husband and I don't know if she will go back. She is the daughter of my son who died of AIDS. She has arrived here very poor, so I am meeting all her needs, and she will have to stay here as there is nowhere for her to go. Where could I send her?

> I have a new granddaughter living here, that one hiding behind the door. She is the daughter of my son who died of AIDS, about eight years ago, but he never told me he had a child.... Her mother married another man and she has been living near here in her mother's father-in-law's house. But now this girl's mother and stepfather have both died and she has come here, to be with her family. I took her in, but they are charging me a goat and a calabash of beer for looking after her – that will cost me about Sh30,000. It makes me sad to remember my son who died, and that he didn't tell me had a child when he was dying. But she came here on her own, and I can't send her away. I must just accept her.

The second largest group of grandchildren being cared for by their grandparents are those who have been left behind by adult children who have left the village. These young people were able to give several explanations for this behaviour:

> Some sons leave their children behind because they see that their parents have a big *shamba* of coffee, and *kisubi* to brew alcohol, and no children to pay school fees for. They think "How can they fail to look after my children well?", and because his parents are rich

he doesn't support them caring for his children at all. There is a proverb: "Why should a calf eat when it expects to get milk from its mother's udder?" (27 year old man)

Who would look after my children when I am at work? (32 year old woman living in Kiguma)

We send the children back to the village because the rooms are very small here and the landlords won't accept children. (30 year old woman in Kiguma)

The above quotes confirm that aged parents are seen as resources which can be exploited by their children when they are themselves facing hardship: life in town is hard, and one's parents can make it easier, even if by doing so, their own well-being is compromised. One elder perceived the problem to be a result of their children having more children than they can care for:

There is a saying: "Produce children that you can look after properly and you will remain well yourself." Sons and daughters should only produce children that they can look after, and only after they have died should a grandparent have to look after his grandchildren. I have three sons; how could I manage to look after all their children and feed them? That would be a great burden!

The relationship between grandparents and their orphaned grandchildren

Radcliffe-Brown (1950) notes that, in many African societies:

Between two proximate generations the relation is normally one of essential inequality, authority, and protective care on the one side, respect and dependence on the other. But between the two generations of grandparents and grandchildren the relation is a contrasting one of friendly familiarity and near equality. (p30)

This quality of relationship was found by Mair (1934) who writes that grandchildren in Buganda could count on "infinite indulgence" from their grandparents, in whose home "severity was not even theoretically expected" (p61). Kilbride and Kilbride (1990) report that in Buganda:

...a close, free relationship obtains between children and grandparents. It is believed that grandparents are likely to spoil their grandchildren. A grandson might even address his grandmother as "my wife" and she would reciprocate with "my husband". (p99)

In Kikole, a grandparent caring for orphans is *in loco parentis*, although the relationship between grandparent and grandchild conforms to that described above rather than to a parent-child relationship. This situation can lead to tension in the home as grandchildren are reluctant to accept the authority of their grandparents. The uses of the term "*mwami*" in the following quotations reveal the relative status of grandparents and grandchildren today:

At my age, a younger person calls me "*mukadde*" [old person], but one of my own age calls me "*mwami*" [husband, sir, or Mr.], to show that we have equal respect for each other.

The mother is responsible for discipline and can punish her child, but a grandmother can't do this. A grandmother may call her grandson "*Mwami*" [husband, sir, or Mr.] or a granddaughter "*Mukyala*" [wife, madam, or Mrs.], so that the grandchild grows up without respect.

Some elders complained that, as a result, they were unable to control their grandchildren, or that their grandchildren abused them. Grandchildren were said to ignore the needs of their grandparents, through stealing food, or failing to fetch water or firewood. However, during group discussions grandparents said that if a child's parents were both dead, the grandparents could assume a parental role and have authority over a child:

> She is my blood and I would bring her up as my own daughter. (William Muwonge EWR=3)

In Box 7.3 A grandmother describes her attempts to discipline an adult grandson:

Box 7.3 Grandparental authority

I had a son who lived in Kiguma, but he died of AIDS, and his son Richard now lives behind me, on the way up to Nabagareka's place. This morning he was fighting with his wife about a little girl who is living with them. She is Richard's child but he had sent her away to live with her aunt in Kinoni. She came back because she was getting beaten badly by her aunt – you can see her arms are swollen – but he wanted to send her back there again. His wife is the girl's mother and wanted her to stay, but he tried to drive the girl away by force and the couple started fighting. When I heard the noise I went up there to sort it out. I found Richard had beaten his wife badly and she had a large bruise on her breast. I asked him why he was beating her, but he just carried on doing it, so I hit him with my stick and gave him a nasty wound on the head, which has been bleeding. He couldn't beat me because I am his grandmother and he has to respect me, and if he did hit me my children would kill him! I have tried several times to get him to change his behaviour, but even though he listens when I speak, when he gets home he ignores everything I said. So I beat him, to punish him.

The relationship between grandparents and the orphans in their care places obligations on both parties. The grandparent expects to help the child marry and achieve independence, and that the grandchild will support him or her in his old age:

> I will have to pay the bride price for the boy, unless we can find out who his father was. But then if we don't find that out I expect the boy will support me in my old age, as he won't know his real father. (Ibrahim Sabiiti EWR=8)

The impact of grandchildren's presence upon caregiver's well-being

Relatively wealthy grandparents were happy to care for grandchildren, who could work and contribute to the household's well-being:

> When you are old and a son or daughter brings you children to help you, you are happy! They are digging in my *kibanja*. (Gilbert Kasibante EWR=5)

> I didn't ask for him, and I don't know why his father sent him, but I accepted him as it meant that they really loved me. If someone gives you a child that is very important, a sign of love. He is old enough to understand everything I say, and if I ask him to bring something he will do it, but he's not old enough to cause me any problems. (Emma Kabenge EWR=2)

If, however, the household is poor, the arrival of orphans presents difficulties, since any surviving parent is unlikely to support the grandparents in caring for these children:

> There have been a lot of changes. Due to shortage of money these days grandparents don't welcome grandchildren whether from a son or a daughter, as they will add to the size of their household. This is because there is a shortage of food, a shortage of sauce, a shortage of everything! (Ibrahim Sabiiti EWR=8)

The traditional preference for the children of sons influenced the degree to which grandparents expended resources on their grandchildren, and their inability to fulfil their caregiving function to the degree they wished was a cause of distress:

> A child of a son costs the family for medical care, food, education etc. But he can come to you when you are very old and unable to provide these things. This can cause you a lot of pain; if you had money you would educate the child, but you don't and there is no-one to help you! You wouldn't trouble so much for a child of a daughter, as she could be taken away by his or her father, or his family, at any time. But the child of your son is like your son; he is permanent. (Henry Ssewanyama EWR=15)

Since a clan is responsible for the welfare or burial of its members, a particular problem arises when the father of an orphan is unknown:

> After she died, my daughter left behind two children who are sick and I don't know who the father is. This is costing me a lot of money.

> A daughter, once she has grown up, just escapes from here, and goes to Kampala. She gets infected with Slim there and when she comes back she dies. We don't know who is the father of her child, and when it dies as well we don't know where to bury it!

Figures 7.1 and 7.2 show a relatively wealthy and relatively poor participant and the orphans for whom they are caring:

Figure 7.1   **A relatively wealthy participant and her four orphaned grandchildren**

Figure 7.2   **One of the poorest participants and her two orphaned grandchildren**

As noted earlier, adult children in urban areas rarely support their parents, even if they are caring for grandchildren:

> Very few sons and daughters come back from town with support for their parents who are caring for their children. Some of them left their children behind just because they didn't want to look after them themselves, and they don't come back with support because if they were concerned about their parents they wouldn't have left their children with them in the first place. A son can be living a good life in town and not want his children to disrupt that. (26 year old daughter of Henry Ssewanyama EWR=15)

A young man said that parents could be asked to care for grandchildren until their deaths, whereupon their father would return home and assume control of the family assets:

> If a son is poor, and unable to look after his children, he will think that his parents will do a better job, so he leaves them in the village and goes to town. Then when his father dies he comes back to take over his father's land and properties. (18 year old, unmarried son of Amos Ndawula EWR=1)

Summary

The AIDS epidemic's influence on older people's lives is multi-faceted. Firstly, as sexually active adults, they are vulnerable to infection. Secondly, as parents, they are required to provide the physical, emotional, and economic resources necessary to care for, and later bury, their children who have AIDS. Thirdly, as grandparents, they are similarly required to care for orphans, to feed, clothe and educate them, and care for them if they are sick. Finally, as dependent old people, they are deprived of any support in their old age that their adult children might have provided. Whilst each of these brings its own difficulties, it is quite possible that two, three, or all four of these eventualities will happen, simultaneously, to the same aged individual.

Young people appear to have no compunction about leaving their children with their parents. Grandparents are bound by cultural norms to accept their role as caregivers of their grandchildren, and consider this a valuable contribution to their community, but are distressed that they are unable to provide what they would regard as a good standard of care. This is an additional stress, on top of the unavoidable grief they must feel following the deaths of their children, and the ongoing sadness that accompanies caring for their deceased children's children.

These difficulties occur within a context of continuing rural poverty and resulting rural-to-urban migration by young people, and produce a decline in older people's well-being, as their resources are consumed by their caregiving activities. For those whose children are ill, dead, or living elsewhere, these resources are not replenished by filial support, and it is unlikely that any support will be provided by other relatives or community members.

Chapter 8

Vulnerable Livelihoods: Coping With the Challenges of Old Age

If I can't work hard, what shall I eat? (Rebecca Nambiru EWR=22)

In preceding chapters I have analysed the livelihoods of the aged people in Kikole, and revealed a wide variation in their ability to cope with the problems they face. I also indicated that this variation has its basis in the diversity found among their individual resource bases: their capabilities, and their tangible and intangible assets. In this chapter I first discuss the concepts of vulnerability and livelihood security. Within the context of the livelihood, I then explore vulnerability among the study participants and examine their strategic responses to the difficulties they faced in attempting to meet the needs of their household members.

The identification of vulnerable populations

Chambers (1988) notes that, historically, there has been confusion between the concepts of poverty and vulnerability, hence the necessity to make a clear distinction between them. To this end, he has put forward a definition of vulnerability:

> [Vulnerability] means not lack or want, but exposure and defencelessness. It has two sides: the external side of exposure to ·shocks, stress and risk, and the internal side of defencelessness, meaning a lack of means to cope without damaging loss. Loss can take many forms – becoming physically weaker, economically impoverished, socially dependent, humiliated, or psychologically harmed. (Chambers 1995 p20)

Davies (1996 pp23/4) notes that "food-insecure vulnerable groups are usually defined according to structural criteria which do not change much from one year to the next". These criteria mostly describe personal (eg age, infirmity, disability) or household (eg female-headed, having a high dependency ratio) characteristics (Toulmin 1992). Swift (1989) defines this kind of vulnerability as "differential" vulnerability while Davies (1996) describes it as "a potential for being at risk" (p23). However, the identification of those who are vulnerable does not allow the prediction of their responses to their insecurity, or facilitate the identification of those among the "potentially" vulnerable who are able to cope least effectively (Davies 1996 p23). The identification of those least able to cope with insecurity can be achieved through livelihood analysis which, as I have shown in previous chapters, allows the exploration and identification of the multiple factors that influence individual and household well-being.

The livelihoods of the study participants

In Figure 8.1 the discussions of previous chapters have been incorporated into a single conceptual framework – that of the livelihood, which I discussed during the introduction to this volume. I have expanded the conceptual framework of Chambers and Conway (1992) by indicating, for each component, the major factors identified by the elders as significant to their livelihoods in Kikole. By exploring the components of Figure 8.1 from a lifecourse perspective, I will illustrate the vulnerability of livelihoods to the consequences of ageing, and to outside influences, in particular the AIDS epidemic. I shall address each livelihood component individually, but each is, as shown, intimately connected with the others, and my comments should be read in this context.

Capabilities

Some aspects of an individual's capability, such as congenital disability, are determined by accident of birth, while others are amenable to purposeful modification and development, through the acquisition of education, experience, and social or political status. Such modification, however, may itself be limited by predetermined, structural factors such as gender, ethnic origin, and the historical social status of one's family. Notwithstanding these limitations, capabilities develop during life, and can be employed in enhancing one's security through the accumulation of tangible and intangible assets. As one reaches old age, however, one's physical and mental powers are likely to decline. At this time the ageing individual attempts to replace these losses in capability through the exploitation of previously accumulated tangible and intangible assets.

Tangible assets

In the cash-short and physically insecure environment of Kikole, most households consider the accumulation of tangible assets in the form of land, animals, savings, stores, and material possessions in general, to be equivalent to the accumulation of cash in a bank or other institution. Such accumulation, therefore, enhances livelihood security, and is considered a high priority by young men preparing to marry, to farm, or to enter business. When younger, some elders had been in paid employment, others had been in business in their own right, or farmed large areas of land and used the proceeds to buy additional land, or to purchase cows or smaller animals, and breed sizeable herds.

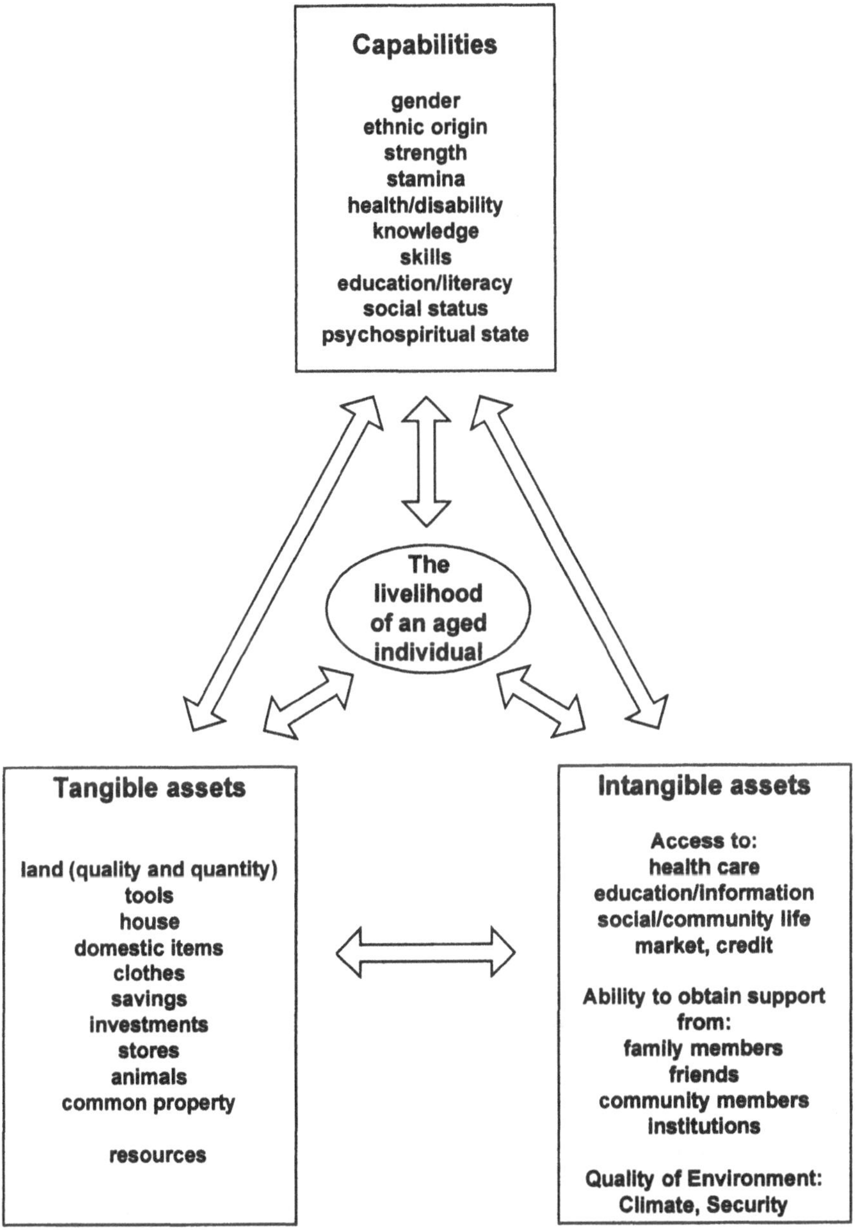

Figure 8.1   Components and flows in the livelihoods of the aged (after Chambers and Conway 1992)

Few are able to continue life in this fashion today, since their declining capabilities are forcing them to consume rather than accumulate these resources. Indeed, some of them are consumed in attempting to slow the decline in their capability, through the use of health services. When they are unable to labour as before, or to engage in business, tangible resources may then be used in paying labour to maintain the production of cash and food crops, but when this is no longer possible land becomes less productive, or is left uncultivated, and may then have to be sold to pay debts or household costs. Cattle die when they sicken and treatment cannot be purchased, and agricultural implements, household items and clothing are not replaced as they wear out. House walls and roofs deteriorate, allowing the elements to enter living quarters. Common property resources such as firewood and wild foods become scarce as the village population increases. Thus, as is the case with their capabilities, individuals' tangible assets are accumulated early in the life course, and consumed later, with detrimental effect on livelihood security.

Intangible assets

In general terms, a child is dependent on its parents not only for its daily needs, but also for the maintenance of its health, its education and its future well-being, whether through a son's inheritance of land, or the advantageous marriage of a daughter. As adults, their reciprocal relationships with their children are supplemented by newly developed relationships with non-family members, and all of these can be called upon for support when difficulties arise. In old age, however, relationships with non-kin become less numerous and less significant, while those with adult children become increasingly important as potential sources of support. Finally, when one becomes frail, one is largely dependent on these adult children. At this time then, one's access to tangible assets, such as food, shelter and clothing, is in fact via one's intangible assets, since it is these relationships that place obligations on children to provide for their dependent parents. A widow, for example, generally lives on and farms land provided by her son.

The accumulation and depletion of resources over the lifecourse

Since, as I have shown, individual stocks of each of these vary throughout the lifecourse, well-being will also vary. To explore the nature of this variation it is first necessary to summarise the changes that occur in individual capabilities and assets throughout the lifecourse. The accumulation and depletion of these resources over the lifecourse is summarised in Figure 8.2.

In Figure 8.2 the left hand axis represents the rate of flow of resources to or from the individual livelihood: the solid line indicates the rate at which outside resources are consumed, while the dotted line shows the rate at which resources are accumulated. If this is above the solid line, then there is a net accumulation taking place, and if below then there is a net depletion. The lifecourse is divided into six periods:

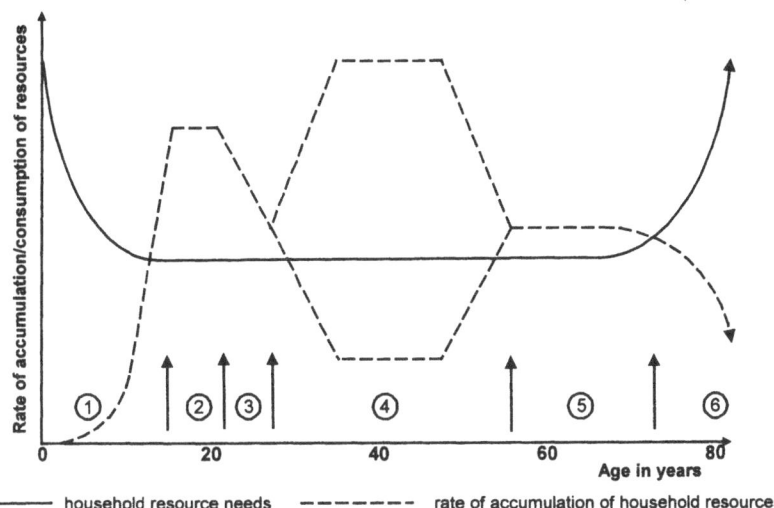

Figure 8.2   The accumulation and depletion of household resources over the lifecourse

1. The period from birth until maturity. The child is at first dependent on others, and gradually becomes able to accumulate resources through its own efforts, achieving independence when the accumulation of resources is equal to the rate of their consumption.

2. Young adulthood. Resources are accumulated at a relatively high rate until children are produced.

3. As successive children are born, more resources are consumed in their care, and the rate of accumulation of resources slows accordingly.

4. The period during which children's ages allow them to attend educational institutions, from primary school to university. Education may be purchased for some, or all, children during this period. The negative impact of this expenditure on resource accumulation will depend on the number of children receiving education, and the cost and duration of their education, while the labour of those that are not in school may make a positive contribution to household resource accumulation. The net change in resource accumulation will follow a trajectory somewhere within the envelope contained in this section. Resource accumulation may increase if many children are not receiving education, but decrease, even to the extent of becoming a net depletion, if many are doing so, or some are attending expensive educational institutions.

5. When all children have left school the household can continue to accumulate resources as long as it is able. During this period, as children are leaving the parental home and establishing their own households, and as their parents' capabilities are starting to decline, household accumulation of assets will at best remain static, but more likely will reduce. At the same time the ageing adults' needs increase, and at the end of this period these increasing needs are seen to be equal to their rate of resource accumulation.

6. During this period the household's ability to accumulate resources is insufficient to meet its needs. Accumulation slows and eventually becomes negative, and its resource base declines since consumption is greater than supply. The well-being of the household now depends on its access to support, and since the period of the lifecourse during which friendships can be called upon for help has passed, children will be the prime source of support.

Figure 8.2 indicates that, assuming one remains healthy throughout one's productive years, there are two periods during the lifecourse where individuals are unable to meet their needs through their current efforts: the periods of childhood and old age. During childhood, needs are met by parents, or other adults, while an aged adult may be able to survive through consuming accumulated resources, or through calling on others for help. A third period, during which children are being raised and educated, is potentially also a time of resource depletion, and unless sufficient reserves have been accumulated help will also be needed at this time.

The ability of a child to support an aged parent will be influenced by the child's circumstances. The earlier discussion on adult children's lives suggests a number of possible scenarios. Some children may be rearing their own families, and the degree of support they can offer can be determined by locating them at a point on Figure 8.2. We know that they will prioritise the needs of their own family above those of their parents, and it is likely that, as their parents become dependent, these children will be embroiled in the task of rearing their own family, and find it difficult to support their parents. Other children migrate to town and rarely visit their parents, who may not know of their child's whereabouts or circumstances. These children rarely offer their parents support, and some of them have, on leaving Kikole, left their children in the care of their parents, thus presenting parents with an additional drain on resources. A final group of children has died, usually from AIDS, and many of these have left their parents with orphaned grandchildren to support. These grandparents then have to look to other children or community members for assistance, which, as I have described earlier, is rarely provided.

Returning to Figure 8.2, it can be predicted that when an aged parent is in the final, dependent period of the lifecourse, it is likely that his or her children will be unable to offer much help, since, if they are in the village, they are likely to be devoting their resources to rearing their own family. By looking further back, to their grandchildren, or, in the case of aged men, to their youngest children, it can be seen that they are likely to be in a lifecourse phase of resource accumulation, and therefore better able to support their grandparents. However, resource accumulation can only take place in an appropriate economic environment, and many grandchildren are moving to urban areas because they find such accumulation impossible in cash poor Kikole. Once they have left the village grandchildren are no more likely to support their aged grandparents than distant children are likely to support their aged parents.

Over the lifecourse, then, an individual accumulates and then consumes tangible and intangible assets. Capabilities can similarly be said to grow and decline as the individual ages. The participants employ all these resources in attempting to meet the challenges that their livelihoods present, and the degree to which this is possible and effective is, therefore, a measure of their vulnerability, or conversely, of their livelihood security. The concept of livelihood security has been described by Davies (1996), and allows for the broadening of this discussion beyond the issue of food insecurity:

> The premise of a livelihood security approach is that food security is a sub-set of needs, neither independent of nor necessarily more important than other aspects of subsistence and survival within poor households. (p18)

The maintenance of livelihood security

Livelihood security in old age is, in large part, determined by the earlier events of one's life. An individual is able, therefore, through strategic actions taken before reaching old age, to influence his or her livelihood security later in life. Through analysis of participants' past experience it is possible to identify strategic decisions aimed at promoting the accumulation and retention of tangible and intangible assets and capability, and thus the livelihood security of their households:

> Life was good when I was strong and could work. It is a result of that work that I am alive today – if I hadn't made a plan and carried it out I wouldn't be here now. (Noel Katusabe EWR=9)

The protection of tangible assets

Income comes primarily from coffee, and therefore it is important to prepare for one's old age by ensuring that land is planted with coffee, and that this coffee is cultivated in order to maintain its production. Other areas of land can then be devoted to seasonal food crops, any surpluses of which may be sold to raise cash. However, to produce well, coffee needs continuous labour input. Aged people had unproductive coffee plantations, or portions of their landholdings that were uncultivated and therefore also unproductive, because they were no longer able to maintain them. Although many aged people had assets, such as cows, that could have been sold to obtain money which could then have been used to pay labourers to restore their land's productivity, they were reluctant to take this action. Their highest priority was to maintain their livelihood security, and avoid any risk of asset depletion, rather than generate income:

It's a risk. What if the crops die, or the coffee does not produce? I have to take a long term view, otherwise I might lose everything! It might even be better to buy another piece of land with my money, and leave that fallow as well, rather than risk the money on seasonal crops. Then you can at least be sure that it will always keep its value. (Vincent Katorogo EWR=18)

In the face of future uncertainty "risk aversion" (Ellis 1988) is thus the priority of most farmers in Kikole. Through leaving tangible assets under- or unemployed, rather than placing them at risk during attempts to maximise profits, farmers are prepared to accept a loss of income, while maintaining or promoting their livelihood security. The values which direct this behaviour are not compatible with either neo-classical or Marxist economic theory, and are the origin of the difficulties that arise when peasant economies are assumed to function, at the household level, according to Marxian precepts. As Scott (1976) observes:

It is perfectly reasonable that the peasant who each season courts hunger and all its consequences should hold a somewhat different opinion of risk-taking than the investor who is gambling "off the top". (p15)

Participants are concerned to maximise livelihood security, rather than their profits, which, as tangible assets, are only one component of their livelihood.

Animals are seen as a type of insurance, as assets to be realised only when there is a specific, and usually unanticipated, need for money. This is not always possible, due to the shortage of cash in the village. The timing of the slaughter and sale of animals is therefore chosen, when possible, with reference to the possibility of being unable to find a customer, the result of which would be a possibly unsustainable loss, rather than in anticipation of profit:

Many others were slaughtering pigs [over Christmas] so we decided not to kill ours. People in the village are poor and we may not have been able to sell it all. (Michael Kavuma EWR=3)

Some wealthier coffee growers were able to maximise profits by keeping their crop until the price was high. This was only possible for those who did not see this storage as a risk, either because of the danger of theft, damage by pests or fungal diseases, or because their immediate need for cash was not so great as to put at risk their household's well-being. Those who were, for example, in need of cash to buy health care, would not risk their future security by failing to obtain health care immediately in order to maximise their coffee profits at a later date. In practice, only those relatively wealthy farmers who had other sources of cash income, and who had secure buildings in which to store their crop, sold their crop at a time of their choosing.

The most extreme demonstration of risk averse behaviour is the rejection of adult children who are deemed a risk rather than an asset. Prodigal, disobedient, or dishonest sons were rejected by their fathers, who were prepared to forgo what they interpreted as a doubtful possibility of future support, an intangible asset, in order to avoid a more likely expenditure of tangible assets to meet their obligations to give them land, to pay a brideprice for them, or to care for their children. Since their own children are aged parents' primary source of support this rejection is not undertaken lightly, and one father rejected a son only after accepting responsibility for two of his children:

> These two children were produced by my son before he was married. Now that he is married he still hasn't settled down, and we don't agree about his behaviour. He could easily leave the next child with me too! I chased him away to Rwabenge, and made a public announcement about it, at a local council meeting, so that I could get peace. He wanted to eat, but he didn't want to work. He wouldn't help me with errands or in the garden, but when it came to meal times he was always there. I still have his children to look after though, because even though I chased him away I couldn't chase them away too.

In Box 8.1 Another participant recounts his treatment of a dishonest son:

Box 8.1 The rejection of a son

I'll tell you what happened. It's a long story! When I separated from my wife my son went with her to her parents' home. I asked her to send him back when he was three years old, and again when he was seven years old but she refused, and he finally came here when he was eighteen. He lived here for some time, but he used to pick and sell my coffee without my knowing, and keep the money. I had a bicycle and radio, only two months old, and he took them and went with them to Mawokota. There he sold parts of them for money, like the mudguard, and the speaker of the radio. The chiefs in that village asked him where he got the bicycle and the radio from, but he didn't have any evidence that they were his. I had a friend, of the same clan as me, who had heard where he was, so we went there together. When we arrived we found that he had sold almost all the parts of the bicycle and of the radio. All that was left was the box of the radio and the frame of the bicycle. I showed the chiefs my receipts and they let me bring them back here, but I left the box of the radio and just brought the bicycle frame. After that I decided to tell that boy not to come to my home any more. Since then he has lived with his mother at his grandfather's home. I heard that he is working in Kiguma now, at a coffee factory, carrying coffee from the vehicle to the machine.

I have described earlier the decision not to obtain health care for a person dying with AIDS. This was another example of risk aversion, in that the potential problems resulting from expenditure on health care, such as a requirement to go into debt, or to deny health care or food to other household members would not be countered by any positive outcome for the patient, who was known to be dying.

The protection of intangible assets

In Kikole, where surplus production is rarely seen, only the wealthiest individuals are able to accumulate resources to enable them to maintain their standard of living in their old age, and, consequently, the concept of "retirement" is irrelevant to the lives of most older people. Rather, an individual continues to labour as long as he or she is physically able, and, when this is no longer possible, hopes to be supported by his or her children. For most participants, therefore, insurance strategies are directed towards developing or preserving one's intangible rather than tangible assets, and specifically the reciprocal relationships one has with one's children. The first priority has to be to keep one's children alive. Infant mortality is high, AIDS is killing many children, and many elders are worried that they may find themselves childless and helpless. Consequently, a father is concerned that his children remain healthy, and in particular, today, that they remain HIV negative.

Since nearby children are more likely to provide help to their parents than those living further away, it is in the parents' interests to encourage their adult children not to leave the village. For sons, the most effective way to do this is to give them land, and pay a brideprice so that they can marry, in the hope that they will build a home and raise a family there. It is less easy to retain daughters, who will go to live with their husband on their marriage. Thus, a local marriage is desirable, or, failing that, marriage to a wealthy husband, whom his wife can persuade to support her parents.

Strategic decision making regarding the passage of land from an older to a younger generation has been noted elsewhere in Africa (Shelton 1972, Colson and Scudder 1981, Glascock 1986, Nugent 1990). Mair (1934) noted the strategic importance of land for the Baganda:

> The best possible safeguard against the future exploitation of the peasants consists in the acquisition of small holdings. (p170)

In Kikole, the productivity of land is said to be decreasing, and therefore more is needed to subsist. Further, as the original few settlers' families have grown through two or three generations, the numbers attempting to subsist on the village's limited land have increased. It is becoming increasingly difficult to persuade children, some of whom are attracted by the prospect of urban living, to remain in the village and support their ageing parents through the cultivation of insufficient or infertile land.

Wolf (1966) observed that where land is in plenty equal division of one's land among one's children, most often sons, is the norm, since, although each may receive less than is needed for subsistence, more land is readily available. However, where land is in short supply, subdivision among a large number of children may make it impossible for any of them to subsist and force them all to leave the village. In this case the unequal subdivision of land is often seen, and the favoured son who receives sufficient land to subsist normally remains in residence, while the others may be forced to leave the village. Although Wolf (1966) asserted that herein lay one of the root causes of rural-to-urban migration and the creation of the urban labour pool which has facilitated economic development, this system would have ensured that at least one child remained near his or her parents, and was thus available to provide support. Southwold (1956) reported that, in Buganda at that time, the heir to a deceased individual normally inherited the largest share of land, and others were compensated by receiving a greater share of the deceased's other possessions. The evidence in Kikole today suggests that a more equal allocation of land is the norm, and this practice may be in part responsible for the high numbers of young people who leave the village in a search for work:

> My father used to show us which piece of land was going to be ours when we were young. But we didn't get ownership until he had given us the authority. Old men are afraid that their children might sell their land if they become owners too soon. Some only give their sons the use of the land, and ownership is only transferred in the will. Sons will be told they can go away to Kampala to try to get rich, leaving the land fallow, but that they should return when they fail to get rich, which is what usually happens, and start to cultivate the land again. (31 year old man)

All male participants, and some female, employed land, their primary tangible asset, in insurance strategies aimed at maintaining their well-being in their old age. During settlement of the village, the male participants, who were young men at the time, purchased land on which to live and rear their families. Since that time they have been able to use the produce of this land to this end, but now that they are growing old some of them are unable to cultivate all their land, and today this land lies fallow, apparently unproductive. However, even before a father's death, his fallow land can be used by his children, and since nearby adult children are the major source of support for frail aged people, participants are keen that this should happen. Thus, fathers were able to insure against destitution in old age by increasing the probability that an adult child would be living nearby at that time. I stress "probability" here, since many children left the village looking for work, and others died of AIDS, and there could be no certainty either that a child would be present to provide the hoped for support, or that, if a child did live nearby he or she would provide support to his or her parents.

Sons who had displeased their fathers, or were seen by their fathers as unlikely to provide them with support, may not be given land, and those who did receive land are usually given only its use until their father's death. Through keeping current the possibility that he would not allow his sons to inherit ownership of land, a father is able further to insure against a lack of support in his old age:

> Some boys just don't think about their futures. A parent may give his son some money and tell him to buy coffee and dry it, sell it and get more money. Then at the end of the week, when he asks that son how much money he has made he will tell lies, saying that he owes money to this other boy and this other boy and then the business collapses. After this has happened a few times – the son has failed – then the father starts to wonder what is the point of giving land to this son if he won't use it? Better to sell the land and drink sugar. (40 year old son of Irene Mutondo EWR=30)

Land is more important as a resource to be passed to the next generation, than it is as a source of income. An old man, now deceased, had no compunction about selling some of his land when his children had abandoned him. This man's wife recalled:

My husband was angry because our children left the village and went to the city and became *bayaye* [wasters]. He realised he was working hard for nothing, so he sold some of his land. (Lilliane Barenzi EWR=27)

Figures 8.3 and 8.4 illustrate the judicious and injudicious division of land among one's descendants. Figure 8.3 shows how Noel Katusabe (EWR=9), who is still alive, has given use, but not ownership, of some of his land to his two sons, Simon and Joseph, but not to any of his six adult daughters. Simon spends most of his time in Kampala, and does not support his father although Katusabe and his wife are caring for two of Simon's children, born outside his marriage. Simon does not cultivate his land, which is now returning to bush. Joseph, who is cultivating about half of his land and leaving the other half to become overgrown, also spends much time away from the village, and as a result his crops are suffering from a lack of attention. In contrast, their father is cultivating almost all his land, apart from a small plot which is fallow as part of a crop rotation. Additionally he is cultivating another half acre which he is renting from a neighbour, in exchange for a share of the crop, if any. He rents from a neighbour while his sons have fallow land because he does not have the power to cultivate fallow land, whereas his neighbour's land had been recently cultivated and relatively little labour was required to plant there. To date, then, his sons have not chosen to remain in the village and cultivate their land, and so he has not gained any increased support through allowing them the use of his land.

Although the adoption of such strategies is considered wise, even if they do not produce the desired result, not all landowners exhibited this risk aversion. It was noted during the introduction, for example, that Rebecca Nambiru's brother, Moses, disposed of much of his land piecemeal, and as a result many of his children inherited so little land that they were forced to leave the village in order to survive. The division of his land is detailed in Figure 8.4, where the disparity between the amount of land originally owned by Moses and that left to his children is clearly visible.

A widow, who is usually living on land which has been left to a son by her husband, is not able to exploit her control of land to influence the place of residence of her children, unless she has acquired ownership of land in her own right. West (1972) notes the potential for disputes between a mother, who may want her property to be passed to her children, and the members of her clan, who may want it to be passed to her patrilineal kin. It may, therefore, prove difficult for a widow to assure her future through land distribution. I found no instances, and no recorded instances, of Kikole women bequeathing land to their children, because there has not been sufficient time since the settlement of the village, which was entirely by men, for this to occur.

The first women to acquire land in the village are still alive, and how they will dispose of their land remains to be seen. A young man told me, during a group discussion, that "a woman always leaves her possessions to her brother's children rather than her own, to keep them in the family clan". Should this practice also apply to a woman's land, her children may not remain near her home, and be available to help her as she ages.

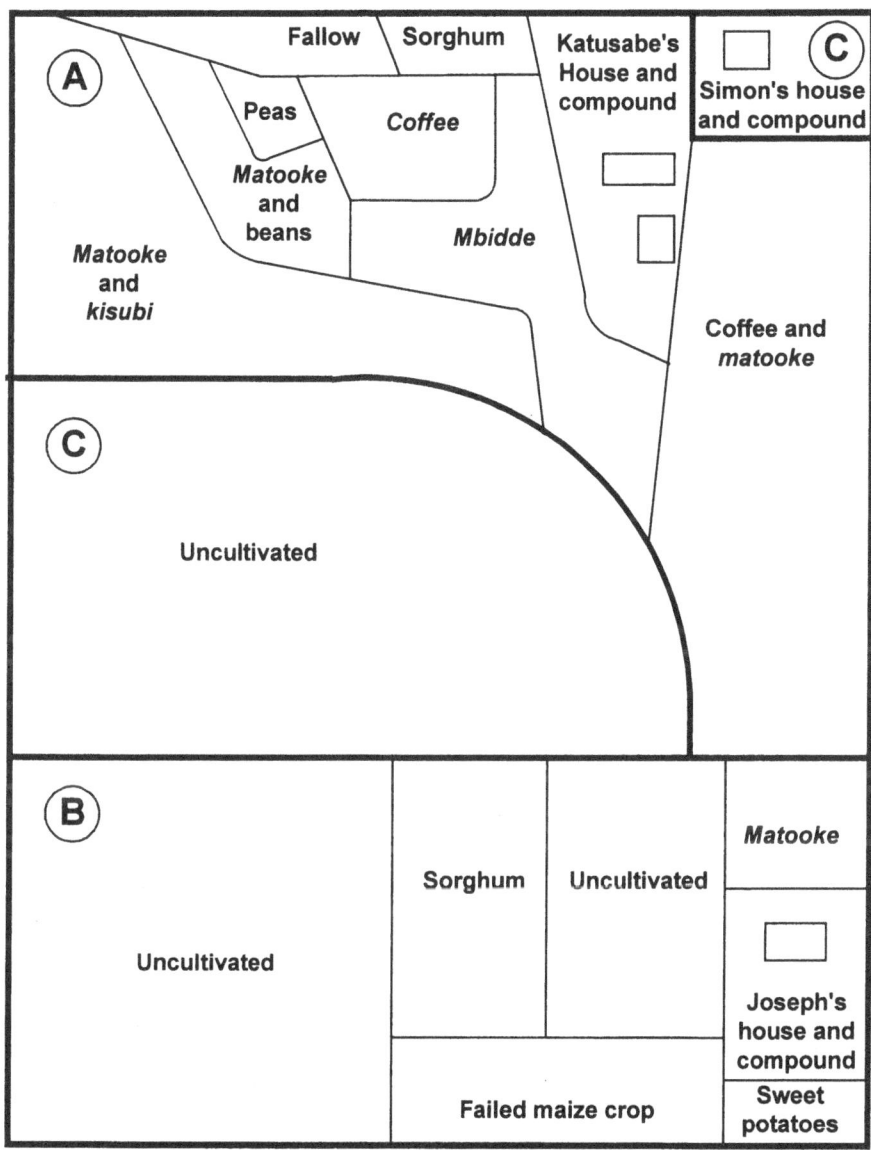

Heavy lines denote the divisions between land being cultivated by Katusabe ("A"), and the land he has allowed his sons Joseph ("B") and Simon ("C") to use. The total area is approximately 10 acres.

Figure 8.3   The division of Noel Katusabe's *kibanja* among his sons (not to scale)

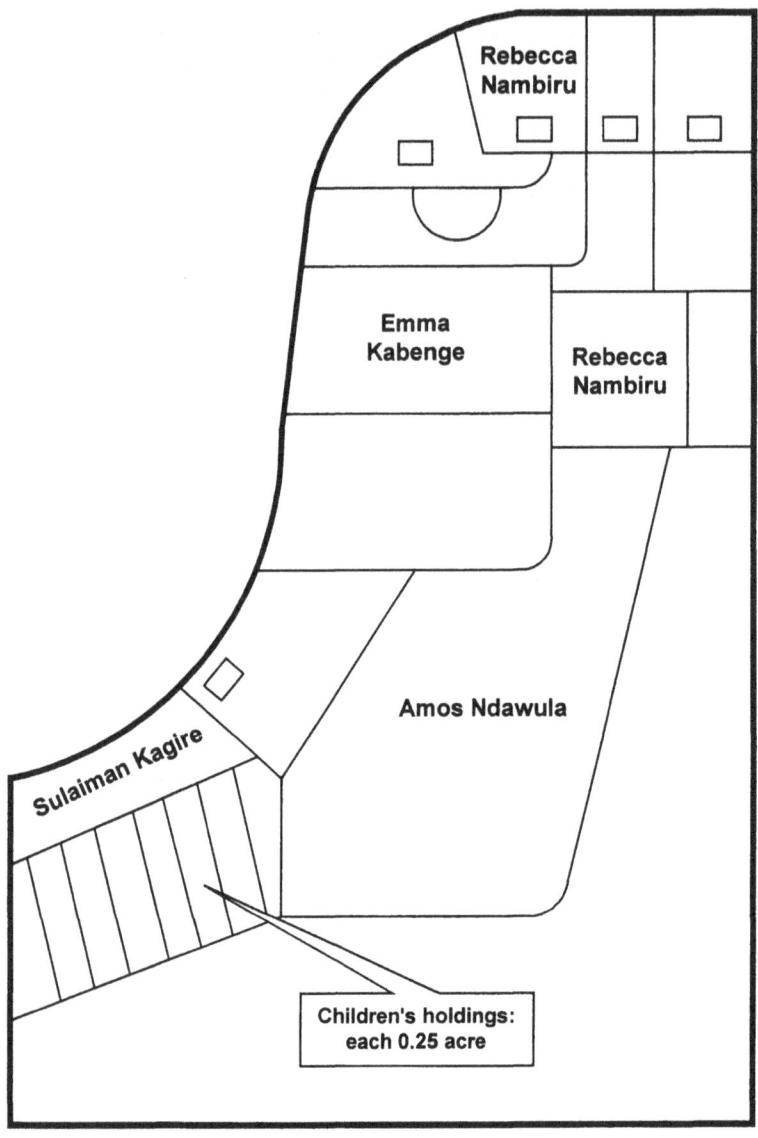

The area of the whole is approximately 30 acres. Many plots were sold by Moses, some to participants in this study, whose names are shown on the land they purchased. The remaining two acres he divided among his eight children, giving them each just a quarter of an acre. All of these children now live in Kampala since they were unable to make a living from their small landholding.

Figure 8.4   The division of Moses's *kibanja* (not to scale)

The education of children was, for some participants a strategy intended to enable them to become wealthy and then provide support to their aged parents:

In the past people used to educate only sons, but now people realise that both girls and boys should be educated in the same way. Then if a daughter gets a good job she can support you in the same way that a son can. (Ibrahim Sabiiti EWR=8)

However, most did not expect this to happen:

Paying school fees for a child is to help him or her have a good future. But when they get good things they forget you... Having children that will support you is just a matter of chance; very few children remember their parents. (Emma Kabenge 043 EWR=2)

Grandparents also accepted the responsibility for educating their adopted grandchildren, although their own low level of education made it difficult for them to know whether their money was well spent:

I don't know anything about education, but I do know that they want Sh3000! (Florence Mugasha EWR=11)

I've paid Sh2000 each for two grandchildren to have extra teaching during the holidays....I don't know whether it's important or not, but the teachers say it is; I just see them going to school and coming home again. (Noah Sserwadda EWR=12)

Mair (1934) observed that, in the early 1930s, education was seen primarily as a means of leaving the village, and, since an absence of employment opportunities now forces young adults to leave the village if they want to utilise their education, this remains true today. As I have shown, distant children generally do not support their parents, and, perhaps unsurprisingly, there was no evidence in Kikole that educating children had improved the well-being of their parents. However, such evidence is difficult to discern, since only a few relatively wealthy parents could afford to educate their children to a standard which would enable them to obtain well-paid employment, or to meet and marry a wealthy partner. Further, these parents are, through their wealth, those least in need of support from their children.

The preservation of capability

Capability has both physical and mental aspects. Physical capability, for the participants, refers to their health, their ability to labour, and therefore to grow food and cash crops. As they age, all of them inevitably experience a decline in their physical capability, and most had both acute and chronic health problems during the study period. The maintenance of health, as a strategy aimed at the maximisation of physical capability, was a goal of all participants, but not all of them were able to achieve this. The circumstances of some prevented them eating well, or maintaining hygienic conditions at home, or obtaining health care when ill, and these participants were therefore unable to optimise their physical productivity. The arrival of AIDS has caused participants to reassess and in some instances to modify, their sexual behaviour. These individuals have, through abstinence or a reduction in the number of their sexual partners, attempted to preserve their capability.

Heavy work was said to reduce lifespan. Those few with money could afford to employ labour to work in their fields, but most could not:

> I look after my coffee by weeding it, but I pay porters to dig it, so that I get a good crop. I haven't the power to dig, and I don't want to overwork myself because I am already weak. If you want to be well in old age you mustn't work hard. (Grace Nanteza EWR=6)

By "mental capability", I am referring broadly to self-esteem and empowerment. Research elsewhere has shown that the maintenance of self-esteem may, on occasions, be valued more than the maintenance of food supply (Beck 1989), and thus more highly than the maintenance of health. Some participants were able to raise their self-esteem through the fulfilment of their roles as parents or grandparents. I have, for example, discussed the importance of being remembered after one's death; the very low wealth ranking given a woman whose children have all died, compared to the high ranking given to her sister who has a single wealthy son; the decisions of some women to remain unmarried in spite of the reduction in their livelihood security that followed this decision; and the satisfaction two relatively wealthy participants obtained from caring for their grandchildren, while others found orphans an unwelcome addition to their households.

Through its passage to the next generation, land is a representation of its owner's success in providing for his or her descendants. Men, in particular, said it was important to fulfil their duty as parents by leaving land to their children, especially to their sons:

> We like to grow coffee because your son will be proud that you worked for him and gave him land with coffee on it. (Ibrahim Sabiiti EWR=8)

Accumulative strategies are, as in Western economies, largely employed during one's productive years, when one has surplus assets to invest in them and the capability to do so. Most of the participants were no longer generating surplus assets, and therefore did not employ these strategies during the study period. Instead, they were progressively consuming their accumulated resources during attempts to maintain their livelihood security in the face of declines in their capabilities and assets. The degree to which these attempts were successful can be taken as an indication of relative livelihood security, or of their relative vulnerability.

Vulnerable livelihoods

A household with sufficient resources, will be better able to respond appropriately to a shock than will one with fewer. Once the shock has occurred and the impact been felt by the household, the household's ability to recover is also a function of its resource base, although not necessarily of the same capabilities or assets as are used to absorb the initial impact of the shock itself. These two dimensions of vulnerability;

the extent to which a system is affected by a shock, and its ability to recover, have been labelled "sensitivity" and "resilience" respectively (Blaikie and Brookfield 1987), and are represented diagrammatically in Figure 8.5.

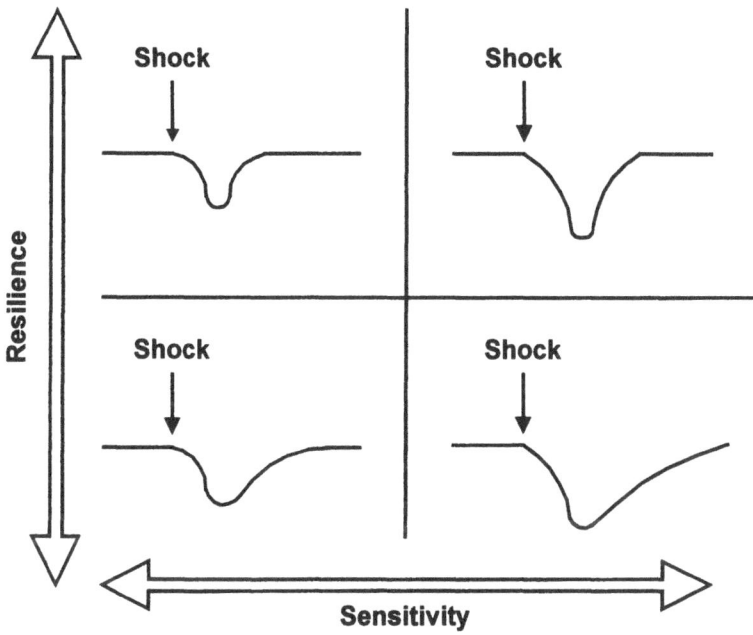

Figure 8.5 ⋅ **The impact of a single shock on sensitive and resilient households (adapted from Davies 1996, p27)**

Figure 8.5 shows that those households which are least affected by a shock are those that exhibit a high resilience and a low sensitivity. The magnitude of the shock's impact is smaller than is the case in more sensitive households, and its duration is shorter than in less resilient households. The figure also shows that those households that are most affected by the shock are those which are the most sensitive and least resilient households.

Figure 8.5 considers response to a single shock. For the study participants, however, shocks, whether they be bouts of ill-health or seasonal crop failures, may occur repeatedly, and possibly before recovery from the previous shock is complete. Some shocks, such as widowhood, the loss of a household member to migration or AIDS, or a reduction in participants' health status as their age increases, result in a permanent reduction in the household's resource base. Figure 8.6 illustrates three patterns of response to successive shocks.

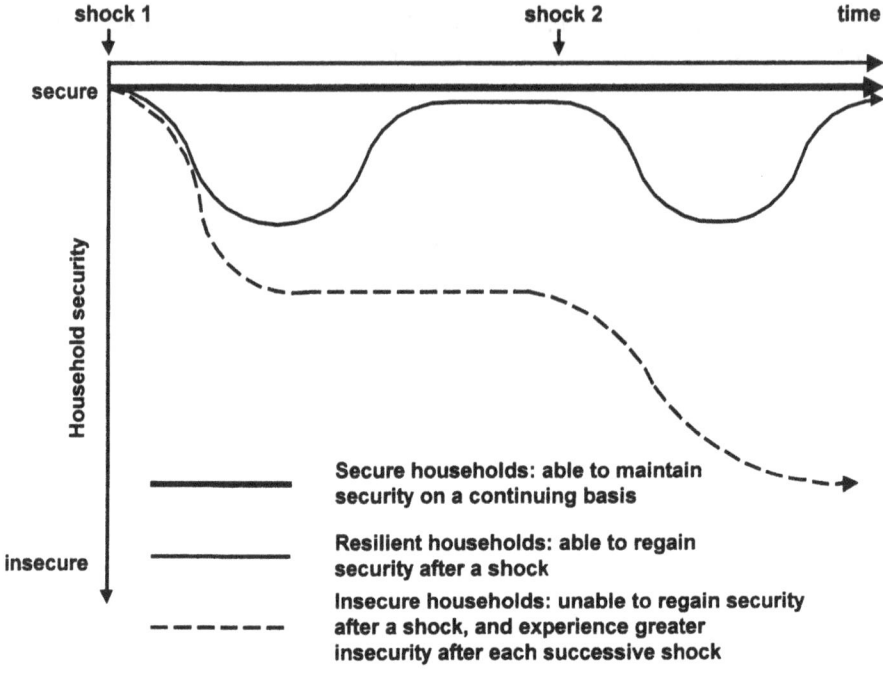

Figure 8.6   The impact of successive shocks on sensitive and resilient households (adapted from Maxwell and Smith 1992, p34)

Households that are most likely to follow the lowest of these trajectories are those with the least resources to call upon. They are the least resilient and most sensitive households, and therefore the most vulnerable. For these households, recovery from a shock requires the consumption of resources which cannot be replaced before the next shock occurs, and, consequently, they are less well equipped to respond to subsequent shocks. They experience a continual decline in their resource base, and therefore increasing vulnerability, after each successive shock. In the following discussion I will describe a household whose livelihood has high resilience and low sensitivity as a "secure" household, and one which exhibits low resilience and high sensitivity as a "vulnerable" household.

Responses to livelihood insecurity

Corbett (1988), in a seminal analysis, asserts that the nature of a household's coping strategies changes as its resource base is reduced following successive food shortages. She places these strategies into three groups, which are used successively in response to these shortages. Insurance strategies involve the consumption of accumulated non-essential assets. They are employed first, with the aim of preserving the primary source of production of the household, which in Kikole, for example, would most often mean the avoidance of the sale of land. Accumulated stores of food, animals, or saleable items are converted to food when a shortage arises, while reciprocal relationships are exploited, debts called in and household consumption reduced, all strategies which will allow the household to survive without the disposal of critical, productive assets. Migration in search of work, with the intention of bringing the proceeds into the household economy, is also an insurance strategy. Clearly, secure households with large amounts of stored resources will be able to cope with longer periods of food shortage, or more severe food shortage, than will those vulnerable households with fewer stores, and the degree to which a household is vulnerable today will therefore be influenced by its past experience of resource accumulation or depletion.

The second group of strategies requires the depletion, through their sale or otherwise, of productive assets, and they thus have more serious consequences for future household well-being than do insurance strategies. It is for this reason that they are employed only after insurance strategies have been exhausted. The most secure households may be able to avoid the sale of productive assets, if they are able to continue to employ insurance strategies until the threat to their security has passed. Productive assets include livestock, tools, and land, and their sale can take place once only. Similarly, there is a limit to the amount of credit an individual can obtain, and once this is reached the resource represented by the ability to obtain credit is no longer available. Further, productive resources that require expenditure for their continuing production, such as animals which need specific foods or veterinary care, may, during a period in which productive assets are being sold, cease to be productive, or die. Through the lack of this expenditure, rather than through their sale, their value to the household may be lost.

The sale of productive assets will compromise a vulnerable household's ability to recover following the ending of a period of insecurity. It may take a long time, or be impossible, to reacquire land, while a herd of cattle could take many years to replace. Vulnerable households that employ these strategies can limit their sensitivity to food shortage, but in doing so they reduce their resilience and are thus less able to recover from this shortage than are those secure households that were able to cope with it through the use only of insurance strategies.

If, once all these strategies have been utilised, the food shortage continues and the household is unable to maintain its food supply, then the final strategy, that of distress migration in search of food, is employed. This action is an indication of extremely high sensitivity to food shortage. Since it may result in the loss of the household's land and home, it may be irreversible, and thus also a demonstration of extremely low resilience. Corbett (1988) notes that in famine situations the numbers of people migrating in this way may be very high, and, without outside intervention there may be significant mortality among this group. Payne (1995) reports that the aged are likely to be over-represented among those who die in this situation.

When available resources are not enough to meet current needs, some needs have to be prioritised ahead of others. The process by which this takes place is of interest, and Corbett's (1988) analysis concludes with the observation that long-term livelihood sustainability remains the primary goal until the subsistence of the household is so seriously at risk that its assurance must take precedence over long-term considerations. As Frankenburger and Goldstein (1990) state:

> The dilemma facing small farm households when coping with threats to household food security involves a trade-off between immediate subsistence and long-term sustainability. (p21)

If this statement is to be related to the lives of the aged, an understanding of the meaning of "long-term sustainability" to an individual in later life, who can expect to experience continually declining capability, has to be obtained. As I have shown, elders were prepared to make personal sacrifices, both financial and material, in order to promote the well-being of their children or grandchildren, actions which they saw as obligatory to the fulfilment of their role as parents or grandparents.

> It is the duty of a grandmother, if she is alive, to look after her son's children until they are grown up. We have a saying: "An elephant has to carry its tusks." (Irene Mutondo EWR=30)

Thus, the above dilemma becomes, for older people, a trade-off between their immediate subsistence and the long-term well-being of their children and grandchildren:

> The money that I would use to help myself has to be used to pay the school fees for these children. We have to give up salt and paraffin and eat by moonlight. (Lilliane Barenzi EWR=27)

> I must ensure that they are educated even if it means reducing what we eat. (Michael Kavuma EWR=3)

Coping and adaptation

Davies (1996), refining the work of Corbett (1988), draws a distinction between coping and adaptive strategies. Coping strategies are those which limit the impact of a specific insecurity through the mobilisation of available resources, following which the remaining resources are employed to return livelihood security to its previous level. Coping is an effective, reversible response to an acute situation, and in doing so, a household is demonstrating a degree of resilience. Adaptation, however, is the response of a non-resilient household. Through making irreversible changes, the household is able only to limit the impact of the insecurity, the degree to which it can do so being an indication of the household's sensitivity. Following its response to this acute insecurity, its resource base is depleted to the point at which it is unable to return to its previous level of security. The acute insecurity therefore becomes chronic for this non-resilient household.

An important distinction between coping and adaptive strategies is that whilst a strategy may be adopted repeatedly as a coping response, when used as an adaptive strategy it can be used only once, after which time its availability as a coping response ceases. For example, a young family may stop eating meat in response to a seasonal shortage of cash, whilst an older head-of-household may take the same decision in response to his or her age-related declining ability to generate cash. The former is a coping strategy, which, in a resilient household, can be reversed when the seasonal shortage of cash has passed, and re-employed when the same circumstances reoccur. The latter is an adaptive strategy of a non-resilient household which, since it is a response to declining capability with ageing, is unlikely to be reversible. Once employed, it cannot be repeated, and the aged individual has one fewer strategies available for employment in the future. Therefore, adaptation, whilst it may successfully deal with the current problem, can be seen, for older people with no prospects of reversing the loss of capabilities or assets which necessitated the adaptation, to result in a reduction in the range of coping or adaptive strategies available for their future use. In this sense, adaptation, which is a manifestation of vulnerability and an inability to cope, can be said, of itself, to exacerbate both of these.

Declining livelihood security among the elders in Kikole

While the use of one's accumulated assets and capabilities to protect livelihood security is essentially a proactive enterprise, involving the voluntary use of available resources, often surplus to current needs, adaptive and coping strategies are reactive, and are undertaken in response to a need, which, for the elders, most often will have arisen through involuntary agency.

In previous chapters I have identified many circumstances which jeopardise livelihood security. Amongst the most significant to the lives of the participants are the loss of capability through declining health, and the loss of tangible assets through the decreasing production of their land, the increased consumption of these assets in meeting the increased demands of their households following the adoption of grandchildren, or of repeated contributions towards the costs of burials. Equally significant are the losses of intangible assets that occur through widowhood, and the loss of children's support following their migration from the village or their death. The majority of participants were unable to prevent a decline in their security as their resources were gradually consumed while attempting to meet the needs of their household members. Of the very few who were able to do so, all but one were relatively young elders; the other was the midwife, who was the only participant to have a regular income.

Since data collection for this study took place over just one year, I am unable to describe the decline of individual households over many years, and can only speculate that as these more secure individuals age they, too, will enter the decline that has been the experience of older participants or their less capable contemporaries. Declining capability has reduced their production of tangible assets:

When I was strong and could dig a lot I used to call myself a real man, and lorries came to pick up all the coffee I grew on my *kibanja*. But now I am old and weak, weak, weak, and the land has become fallow. I used to be able to get money to buy clothes for my wife, to pay for health care, and my graduated tax, but I can't meet all those needs today. (Vincent Katorogo EWR=18)

Intangible assets, here in the form of social status, were also threatened by advancing age:

I used to be a rich man, but since I lost my wife my life has become very poor. I used to be able to dig, but these days I can't do anything. Look at my clothes! I used to be a chief, but if people see me like this they will just laugh at me. (George Kabongo EWR=16)

The other day my granddaughter told me to go and dig! It was painful for me to hear that, as she wouldn't have said it when I was younger. I'm just waiting to die, all the others of my age are resting in peace and I don't see what I'm waiting for. (Edith Manube EWR=29)

I am no longer a human being! My tongue is like a leaf blown by the wind and I say whatever it wants.... (Ester Namukisa EWR=27)

Adaption to changes in livelihood security, through an irreversible consumption of available resources, is said by elders to result in a decline in household well-being, a decline which must be taken as evidence of an inability to cope effectively. I argue, therefore, that among the elders of Kikole, once declining capability has a significant effect on one's productivity, a qualification which excludes those younger or wealthier participants whose livelihoods are secure at present, failure to cope effectively is the norm. For them, adaptation to each, successive, shock produces a further decline in household resources until existence is at the level of simple subsistence.

Irreversible adaptive strategies, since they are a function of available resources, vary between households, and it is not possible to predict individuals' responses to specific problems. However, the range of available responses is a function of the range of available resources, and therefore contracts after each livelihood adaptation. Thus, those who have been forced to consume resources in successive adaptations find themselves increasingly constrained in their responses to subsequent threats to their security, and those who have the fewest remaining resources will be least able to cope with these threats. Referring, once again, to Figure 8.5, this decline can be represented as a move from the upper left to the lower right quadrant; that is a shift from a livelihood exhibiting high resilience and low sensitivity, to one of low resilience and high sensitivity. Whether the pathway taken by an individual livelihood during this transition is directly across the centre of the diagram, or via the upper right or lower left quadrants, will be a function of their specific and changing circumstances. Since it is in part a function of the declining capability that accompanies advancing age, this decline must be age-related, as is shown by the age-related decline in relative wealth shown in Table 8.1:

Table 8.1   Mean emic wealth ranking of elders in Kikole by self-reported age

Self-reported age group		Mean emic wealth rank
60-69	(n=14)	11.6
70-79	(n=12)	16.9
80+	(n=4)	24.5
Total	(n=30)	15.4

Seasonal variation income or in the availability of food adds another problem to those with few remaining resources:

> I have no income apart from my coffee. At this time, at the start of the season, I feel like a rich man, but at the end I will be poor again. (Gilbert Kasibante EWR=5)

> Ha! These days we don't have money! We peasants are poor at this time since it is not the coffee season. (Henry Ssewanyama EWR=15)

Whilst income is confined to two coffee seasons per year, money is needed throughout the year:

> I've had no income recently because it's not the season for coffee and the bananas aren't ripe. That's why I didn't go for treatment when I was sick. (Lilliane Barenzi EWR=27)

> I'm ill because I am worrying so much. My grandchildren are always wanting school fees, clothes, and food, and then I have my own needs. I haven't had treatment because I have no money. (Rebecca Nambiru EWR=22)

At the level of subsistence the household remains with minimal tangible assets and relies on the exploitation of its remaining capability and intangible assets. If the aged person is capable, he or she can perform wage labour. Credit may be obtained from shopkeepers or coffee dealers, the availability of support from friends having declined along with one's tangible resources.

Wage labouring as a coping strategy

> I may work four days for someone and then two days on my land, before working for someone else again. It depends on the problems I have at the time. If I need money for school fees then I work more for other people. (Rebecca Nambiru EWR=22)

As a coping strategy, wage labouring is problematic because one cannot be sure of finding work when one is in need. Further, when work is to be found, during the season when an employer's land is ready to be dug, then it is likely that one's own land is in a similar condition, and by working for money one could be endangering one's domestic food production. A balance has to be struck between labouring for money and cultivating one's *kibanja*:

> Last season I didn't harvest anything. I was only working for money for the family, the work was very hard and it exhausted me so that I couldn't work on my land. And now this season, because I was working for money, I have planted my crops late. (Solomon Nsamba EWR=13)

Thus, even at the level of extreme poverty that forces one to work for others, such a coping strategy further compromises one's future livelihood security. Working in this way is qualitatively different from other income generating activities, which can be carried out in addition to working on one's *kibanja*:

> Working in small jobs like brewing or selling bananas, or repairing bikes isn't the same as working on others' land because they don't prevent you working on your own *kibanja*. Those jobs can be done later in the day, so you can still grow your own food and coffee. Digging for others is done instead of working on your own *kibanja* and has a bad effect on your life. (Tax collector)

However, these alternatives are insurance strategies, and all require the investment of resources. They are therefore beyond the means of those who are looking for work because their tangible assets are exhausted.

Borrowing as a coping strategy

Most elders were in debt at some point in the year, and some were in debt throughout the study period. Borrowing money was seen as a risky strategy and those who were not forced to do so avoided the practice:

> We do without. If we get credit then we have a lot to pay and have nothing in reserve. If you go on increasing your debt then the shopkeeper can evict you from your land to get his money! (Noel Katusabe EWR=9)

In order to obtain credit one has to be seen as creditworthy in the eyes of one's potential creditor, and the borrower has to convince the lender that he or she has enough assets to act as security for the loan required. For wealthier farmers, with land and cows in quantity, this is not a problem, but the majority of participants are not in this position. One man, who was not among the poorest participants, was unable to obtain credit due to his having failed to repay previous debts:

> We sometimes go four days without paraffin, and on the fifth I will approach a shopkeeper with lies about how I will take him money very soon, and then he will give me what I need. But these days people are refusing absolutely to give me credit!

The creditworthiness of poor participants is extremely limited, and therefore the sums involved are usually small. Once such loans had been obtained no more were available until the earlier loans had been repaid. However, difficulties arose in repaying even small sums:

> I had to use the money for school fees. As soon as I have paid the Sh10,000 to the school I will pay him back. (Rebecca Nambiru EWR=22)

Later, Rebecca had to delay repayment again:

> I had to use the money because I had toothache. I will pay him eventually from my coffee, when it is ripe. (Rebecca Nambiru EWR=22)

Another man, who had been forced to borrow to pay a debt incurred by a granddaughter, prioritised repayment even above his own health:

> You won't see me putting coffee out to dry because I have to sell my coffee straight away, fresh, to repay my debts......I can't go to hospital for health care until I have paid these debts. (Noah Sserwadda EWR=12)

Small amounts of credit were advanced, interest free, by village shopkeepers. Since they were running a business amongst a cash-short population, they had little option but to do so. For larger expenses such as school fees and health care costs, more formal arrangements are entered into with wealthy village residents, shopkeepers, or coffee dealers, involving written contracts drawn up and signed by both parties, and witnessed by a third party. Illiterate villagers are open to exploitation when entering into these arrangements.

Coffee dealers, who have access to ready cash, are frequently approached for larger amounts of credit, and they will often choose to express a debt as a number of tins of coffee rather than a cash sum. Since coffee dealers are not members of the village community, they do not have to concern themselves with its moral economy. Villagers are vulnerable to exploitation in this situation, as they may not know the current price of coffee, and cannot predict the future price. It is likely that any market price changes will work to the advantage of the creditor who has knowledge of them, and against the interests of the debtor who does not.

A poor season could mean that a debt may not be repaid and will have to be carried over until the next season. Friends may accept this situation but businessmen who have lent money may insist on payment, forcing the sale of assets or the taking out of a fresh loan to repay the first. If the creditor is prepared to wait, the defaulters are obliged to sell the following season's coffee as soon as possible to repay the debts. However, selling coffee to clear debts at the start of the season means that the coffee will be sold at a lower price than could be obtained later. This will add to the cost of a loan that was expressed as a number of tins of coffee rather than a cash amount. Similarly, at the start of the season coffee may be sold fresh in order to settle cash debts quickly, reducing the income that could be obtained from the coffee through drying it and keeping it until a better price was offered. Credit, then, while useful in helping the poorest participants to cope with specific problems, was potentially a source of greater insecurity, should the debtor's precarious circumstances prevent payment on time.

Eventually, one's capability declines to the point at which one is no longer employable, and credit lines become exploited to the full. At this time, the only options open to those who are not in receipt of support from others are distress migration, which Corbett (1988) describes as the last resort of the destitute, the sale of productive assets, or begging. The first of these options is not open to frail old people (Gould 1988), while the second is of limited use. By the time an individual has reached this level of deprivation it is unlikely he or she has any animals alive, and the only assets that could be sold are more likely to be small material assets or land:

> An old person who was rich at one time can sell his assets in his old age, say a table, or a cupboard. And if he cannot move, he can send a child to get drugs for him so that he can get treatment. But if his relatives don't support him he must sell his doors, and windows, all his property, so that he can have a good life. Some old people have sold part or all of their land; there are many in that position around here. (57 year old man)

One of the poorest participants, in an act which must have further compromised the security of her livelihood, had sold some land to buy clothing:

> I sold a small piece of my land about two years ago to my neighbour who lives in the house with the glass windows. I was very poor and had no clothes to wear. I bought a *gomesi* with the money. (Juliet Kisaakye EWR=25)

Those with children living nearby are unlikely to have to take such extreme action, since the children would be reluctant to allow the sale of productive assets which they would otherwise inherit. Those without children nearby, and those who have exhausted their stocks of productive assets have to rely on the charity of others for support. However, at a time of life when such reciprocal relationships have been shown to be in decline, or even non-existent, subsistence may become extremely difficult or even impossible. I found only one instance of survival through begging: a very old widow, who was living on land that had previously been owned by her husband, and was now owned by her grandson. She had no land to sell, and relied on passers-by for most of her food:

> I don't look after myself. Other people do that: passers-by! One may give me sugar, money, salt, whatever they can. (Ester Namukisa EWR=27)

This very old woman died during the study. Another very old and very poor widow, also living on her husband's land, had resorted to begging in the past. She lived next-door to her daughter and son-in-law, but they were heavy drinkers and did not provide support on a regular basis. Her home was hidden from the path through the village, and she had to walk through the village when begging. Now that her strength was failing she was no longer able to do this:

I used to move around the village and kind people would give me food but now my legs can't take me that far. There is no more I can do; I will sit here until I die. (Edith Manube EWR=29)

Summary

Chambers (1995), analysing deprivation, identifies its dimensions as vulnerability, poverty, social inferiority, isolation, physical weakness, seasonality, powerlessness and humiliation. Even relatively secure participants in Kikole experience some of these, although they were able to do so without compromising their livelihood security. As participants experienced the decline in their well-being described above, the breadth of their deprivation expanded to include more of these dimensions, although the course of this expansion necessarily varied with their circumstances. By the time they reached the level of subsistence it was apparent that they were experiencing all the dimensions of deprivation listed by Chambers. However, even in such desperate straits, ageing people are bound to continue to experience declines in their capability, and their subsistence cannot be sustained indefinitely. Death in circumstances of extreme poverty appears to be the likely fate of all but the wealthiest villagers.

Chapter 9

Sustainable Livelihoods for the Aged?

The vulnerability of nearly 90% of Uganda's population to both natural and economic forces has the effect of perpetuating the socioeconomic decay of the countryside and rendering non-formal social support systems increasingly more fragile – thus leaving the population without any meaningful form of social insurance. (Ouma 1995 p8)

The present day circumstances of participants of the study are greatly influenced by past experience. Almost all have lived within the constraints set by the quality and quantity of the land they were able to purchase when they first settled in the village. Literacy has enabled a few to transcend these limitations, to accumulate resources and become wealthy. The well-being of most has, however, been compromised by lifecourse events such as illness or disability, marital separation, widowhood, the deaths of children, or the theft of property. The effects of exposure to undernutrition, inability to obtain appropriate health care, poor clothing, housing, bedding, and limited access to water and sanitation facilities, cannot be measured in a study such as that reported here, but each of these can be assumed to have an additional, negative effect on well-being when experienced over a long period.

Every individual has a unique life experience, and accumulates a unique portfolio of tangible and intangible assets which are available for consumption during his or her old age. As these are consumed, livelihood insecurity increases. Among any aged cohort there will be those, probably younger and healthier, currently experiencing secure livelihoods; others, most often older and frailer, who are extremely insecure. Since only the wealthiest can expect to die before their available resources are consumed, most will die after a period of livelihood insecurity. In Kikole this is experienced most frequently as an inability to maintain adequate nutritional intake or obtain health care. Livelihood analysis reveals a high degree of diversity among the aged, but this exists, largely, not in a variation among individual livelihoods, but in the observation of multiple, similar, livelihoods at differing stages of their common experience, that is, their decline in old age. Thus, the security of individual livelihoods will change with advancing years, even though the aged population cohort will continue to be composed of individuals with livelihoods of high and of low security. Those who die after a period of insecurity are replaced by younger individuals, entering the cohort at a time when their security is relatively high.

Livelihood analysis and the study of coping strategies reveals that although, at any one time, a minority of the aged population in Kikole are experiencing the highest levels of insecurity, almost all of them will have this experience before they die. Among elders in Kikole, death in deprived circumstances is the norm rather than the exception, and the description, contained in the previous chapters, of the lives and the coping strategies employed by the poorest participants today, will be applicable to all but a few of the others, as their resources are consumed in coming years.

Livelihood analysis has provided an invaluable framework for the detailed description of the lives presented in previous chapters. However, this technique takes no account of lifecourse-related changes in livelihoods. These changes are the consequences of ageing within an impoverished social and economic environment, and encompass both non-age related changes that result from external agency such as crop failure or burglary, and those that arise in the ageing process itself, such as decreasing capability, the loss of a spouse, or the forced adoption of grandchildren. In the exploration of the lives of the aged, an awareness of their lifecourse experience is central to understanding their contemporary circumstances.

Among the participants, both material and social factors mediated quality of life. Whilst almost all existed in conditions of physical deprivation, the negative impact of these was for some, and to some extent only, mediated by the satisfaction they obtained through fulfilling their roles as parents and grandparents. For example, male participants gained pleasure from passing productive land to their sons, and leaving many children to ensure the future welfare of their clan, while aged women enjoyed their successes in having reared children to adulthood in extremely difficult circumstances, and in their ongoing relationships with these children and their children.

Powerlessness, one of the dimensions of deprivation listed by Chambers (1995), is a major obstacle to the maintenance of quality of life among the aged. The sadness which, in previous chapters, has pervaded so many of the participants' statements, arises from their powerlessness, and in particular from their inability to prevent the loss of their children to childhood illness or to AIDS. This powerlessness can be seen in other areas of their experience. They are unable to prevent the declines in capability that accompany their ageing, or to control the economic environment in which they attempt to accumulate or retain their tangible assets. Inequitable access to markets results in sub-optimal income, with negative consequences on their ability to obtain, for example, health care for themselves or their children, or to make investments to maintain the production of their land, production which is already falling as their capability declines. Similarly, in the context of the limitations of the rural economy, they are unable to prevent the migration of their children to urban areas, with consequent losses to their intangible assets. These losses are magnified by the further losses of children to AIDS, and the acquisition of orphans which are, effectively, negative intangible assets, or liabilities.

At the level of deprivation experienced by the most insecure participants, the use of the word "coping" has to be reconsidered. Certainly, they are able to sustain life by, for example, reducing the quality and quantity of their diet, tolerating insect pests and a low level of personal hygiene, or allowing their homes to collapse around them, but the employment of such strategies is an expression of their powerlessness rather than of assertive, considered decision making, and since it serves only to increase their livelihood insecurity, "coping" cannot be said to be taking place.

The literature concerning coping strategies is comparable to that concerning the livelihood in that it has not, to date, encompassed factors related to the lifecourse. The assumption in the coping strategy literature is that if resources are available, or made available to an individual, then his or her livelihood will remain secure, or, if currently insecure, will regain security. This is clearly not the case among many of the participants in Kikole, whose livelihoods are in decline. Additional resources might be able to slow, but will not reverse, this trend. Discussing coping strategies in situations of food insecurity, Davies (1993) comments:

> Groups who have fallen out of the bottom of livelihood systems are uniquely vulnerable and indeed have to cope to survive. But it is conceptually confusing to lump the means of subsistence eked by such people (the ultra-poor and the destitute) with pre-planned strategies used by people within a livelihood system to overcome an exceptionally severe period of food insecurity. Further, searching for and monitoring coping strategies can mask the collapse of livelihood systems by pre-supposing that people cope even in subsistence economies which are no longer viable from the point of view of either food or livelihood security. (p61)

Implicit in this statement is that the identification of coping strategies and the facilitation of their ongoing utilisation is insufficient to meet the needs of the most vulnerable, and that, in the absence of a lifecourse perspective, the support of coping strategies may well do no more than maintain their state of high insecurity. Just as, by focusing on livelihoods, one may fail to take note of the limitations that lifecourse changes place on the ability of the aged to sustain their livelihoods in spite of the resources available to them, so too a focus on coping strategies may also disguise their changing ability to cope over the lifecourse.

Interventions to support the aged

Interventions in support of the aged in developing countries are few. To date, the numbers of the aged have been low, and their relative invisibility may have resulted in little interest being taken in their welfare. They will increase dramatically in coming decades, however. Some authors and some governments maintain the view that interventions are not needed, as the extended family will continue to care for its aged members. But, as Treas and Logue (1986) point out, when resources are short, individual aged people or their families may be forced to prioritise the well-being of the young over the old. This attitude will not foster an environment likely to encourage the development of supportive interventions. Others, with some justification, cite a lack of resources for the failure to provide state support (eg Hashimoto and Kendig 1992). The resources of these countries are few, their needs are many, and in such as situation:

> ...the most fundamental dilemma [for governments] is whether or not scarce resources should be devoted to a group which is not viewed as an investment in human capital for national economic development. (Hashimoto and Kendig 1992 p6)

The concept of the "sustainable livelihood", as a goal for interventions, has been growing in currency since the late 1980s. Chambers and Conway (1992) define livelihood sustainability as follows:

> A livelihood is sustainable which can cope with and recover from stress and shocks, maintain or enhance its capabilities and assets, and provide sustainable livelihood opportunities for the next generation. (pp 7/8)

Enhancement of livelihood sustainability is a desirable outcome of an intervention because it implies independence from institutional support. Therefore, such interventions are likely to be preferred by governments and others seeking to support the aged in a cash-short environment. However, most of the livelihoods of the older people I have described are not sustainable according to this definition; they cannot cope with stresses and shocks, their land is losing its fertility, and many of their children are leaving the village, usually for an urban life of near-destitution. Most importantly, the livelihoods of the aged are, in the final analysis, unsustainable since they are approaching the end of life. For the aged in Kikole, fulfilment in old age is obtained, not necessarily through the extension of life, but through the maintenance of its quality. If quality is maintained then sustainability is desired, but the powerlessness renders most unable to prevent the declines recounted throughout this volume, and there comes a level of deprivation beyond which, as several participants intimated, the continuation of life becomes undesirable.

It is possible to describe three groups of the aged. Firstly, there are those younger, more active aged people who are likely to be able to act to sustain, if not improve, their circumstances. Second is a group whose members' capability is in an early stage of decline, and who are maintaining their quality of life, at the expense of their livelihood security, through the gradual consumption of resources accumulated throughout their lives. A rapid assessment is unlikely to distinguish this group from the first, since they differ only in their stocks of resources, which are unlikely to be revealed during a cursory inspection. Members of these two groups will be found in phase 5 of the lifecourse, as shown in Figure 8.2. Only when these resources are exhausted will members of this group easily be identified from members of the first, as they will then join the third group, which consists of individuals whose resources have been exhausted and who are dependent on others for their support. Members of this group are unable to influence their circumstances due to their poor physical or mental condition, and their lack of resources. They are in phase 6 of the lifecourse, as shown in Figure 8.2.

In view of the existence of these distinct groups, those planning interventions directed at improving the sustainability of livelihoods of the aged must take account of the fact that while their needs may be similar, their capabilities and assets certainly differ. An awareness and consideration of probable future lifecourse events is essential during the development of any interventions intended to support the aged. It can, for example, be confidently predicted that physical ability will decline with advancing age, and any supportive intervention which relies on a continuing high level physical activity for its success is unlikely to succeed, for any specific individual, for more than a few years.

Chambers (1995) notes that "a livelihood cannot be sustained if its main asset....the body.....is sick, damaged or disabled" (p38), and advocates the use of a "safety net", as proposed in the 1990 World Development Report (World Bank 1990), to protect the vulnerable poor from destitution. Chambers here acknowledges that the declining physical or mental condition of the aged is likely to render their livelihoods unsustainable. The UNDP, which has adopted the sustainable livelihoods concept as "an overarching normative goal for development programming" (Singh and Lawrence 1997 p1), considers one of its most important strengths to be its focus on empowerment rather than welfare. However, the UNDP also acknowledges the need for safety nets for those for whom empowerment is not a realistic goal, to prevent them falling into destitution, and included their provision among the strategies for poverty reduction contained in the *Human Development Report 1997* (UNDP 1997).

Once offered to those in difficulty, welfare cannot be withdrawn without harming the recipient unless the difficulty has been resolved. In the case of the aged, however, whose problems are likely to increase rather than decrease in severity, welfare will, most probably, be needed until the death of the recipient. From the point of view of cash-short governments or NGOs, it is therefore, desirable to delay the provision of such support as long as possible, through the support of self-sustaining empowering interventions, directed at individuals who are not yet dependent on others.

In developed countries today and for the past 20-30 years, pressure groups (the Grey Panthers in the USA, the Australian Pensioners' and Superannuants' Federation) have been countering the marginalisation of elders through maintaining a visible presence in the political arena (Friedan 1994). In developing countries elders have no such advocates and are widely seen as unproductive. Donor countries are reluctant to see funds expended on meeting the needs of the aged in developing countries. Programmes rarely include the elderly, who, as a result, are marginalised, and unable to enjoy the benefits that development makes available to others (Neysmith and Edwardh 1984, Neysmith 1990). Empowerment for the aged people in Kikole does not, however, currently involve issues of retirement or pensions, which have little relevance to the life of any rural Muganda. More important to their lives is the ability to fulfil their roles as parents and grandparents, while maintaining the security of their livelihoods, thus ensuring that their well-being remains at an acceptable level until death.

Supportive organisations, acting as advocates for the aged, such as the Uganda Reach the Aged Association (URAA) have come into being relatively recently, are minimally resourced and as yet inexperienced. The URAA is reliant on external support from HelpAge International in the UK for its continuing existence, and although its work is valuable it is also limited. Few of Uganda's aged population would, as yet, be aware of its existence, just as few younger people have experience of working with advocacy organisations, or would know where to obtain such experience. The small number and limited activities of organisations supporting the aged in developing countries renders the drawing of generalisations concerning their activities problematic and of questionable value. Tout (1989) contains a review and critique of those existing in the late 1980s, amongst which are examples of those developed with and without community participation. Tout (1989) stresses the need to encourage participation, and proposes appropriate guidelines for future interventions.

Participatory research has, as its goal, the empowerment of those taking part (Chambers 1994). Robinson-Pant (1996) considers the individual empowerment resulting from PRA activities to be comparable to that obtained through the achievement of literacy, while Chambers (1997) asserts that participatory research is, inherently, an empowering experience for members of marginalised groups:

> It may be necessary to find those who are excluded, and to bring them into a participatory process, or help them to generate their own.....In participatory processes the challenge is to sequence and balance a coming together around common interests, and recognise and support diversity, complexity and multiple realities to empower those who are weaker and excluded. (p187)

Participatory rural appraisal with the aged has been facilitated by international NGOs such as HelpAge International (eg Thomas and Atkinson 1995) and *Redd Barna* (eg Guijit *et al.* 1994). These appraisals have produced community analyses of the problems elders experience and estimations of the relative importance of these problems. These can then be used as indications of priorities to be addressed by local organisations. Since there was no such intervention in Kikole, I will describe a URAA programme in rural Eastern Uganda, as an illustration of the value, to the aged community members, of an empowering intervention.

URAA assisted older village residents to establish a cattle and pig breeding venture. The animals were owned by an association of which the aged residents were the only members. These members had elected a committee from among their number, and which was charged with the administration of the association. Their animals were kept in shelters, and fed with grasses and other materials grown or provided by the members, and transported, by them, from their homes. Milk from the cows and profits resulting from the sale of animals were distributed among the members, while gas produced by fermenting animal wastes was used to cook a midday meal for the members. Started in 1991, this venture is now both self-sustaining and profitable.

Not all aged people were capable of participation in these empowering interventions, as became clear when I explored the membership of the URAA intervention in more detail. After a wealth-ranking exercise with all 26 elders of the village, I was able to show that, although membership was open to all, the members of this co-operative venture were found almost exclusively in the middle of this ranking. Individuals near the top of the ranking were not taking part, while seven of the bottom ten were excluded. Those at the bottom of the ranking were the most frail of the aged community members, and thus unable to participate in the group's activities. The members themselves, however, benefited from their activities, and, as would be the case in Kikole, used their increased resources to support their families rather than those older people who were unable to participate. Therefore, while those taking part experienced an increase in their livelihood sustainability, there remained a significant proportion of the aged population whose needs were not addressed by this empowering intervention. The needs of this group, and of similarly disadvantaged aged people in Kikole, can be met only by supportive, welfarist action.

If empowering interventions on behalf of the aged in developing countries are few, welfarist interventions, at least those undertaken by governments, are almost absent, and those that do exist are unlikely to be available to the rural elderly (Gibson 1992). Social security benefits in developing countries are generally few, meagre, and available only to a small proportion of the population, amongst whom civil servants and government employees are found to be over-represented (Ogawa 1992). In Uganda, for example, the benefits of the National Social Security Fund are available only to the less than ten per cent of the population in formal employment and whom are able, through their employment, to contribute to a pension fund. Further, these few are almost exclusively urban, and thus the rural population is left without any access to state support (Ouma 1995).

In spite of the paucity of pension provisions for elders in developing countries encouraging emerging research findings suggest that non-contributory schemes which provide the very poor with small, regular payments, and which place little strain on national budgets, can have significant benefits for elders and their families (Devereux 2001, Heslop 2002a).

Governments frequently justify their failure to provide welfare to the aged on the grounds that the extended family fulfil this role (Phillips 1990). In Kikole, these grounds are not valid. In the absence of government and family support, welfare is supplied to some by NGOs, and in Uganda there are national and international organisations offering support to those in need. Many provide education or health care, but few support the aged. These organisations' offices line the tarred road south from Kampala, through Kiguma to Masaka, but their services are not made available to those who do not live in such relatively favoured locations. In Kikole, the only NGO in the area provided free education to those children who were prepared to abandon their Roman Catholic faith and become Baptists. While this service was of help to some of those participants caring for children, those too frail to care for children, or who were themselves being cared for by a grandchild, did not benefit from this NGO's activities.

The current concern with the welfare of AIDS orphans illustrates a widespread attitude among academics and aid agencies to the problems of the aged. In sub-Saharan African countries, the multiple impacts of the AIDS epidemic on the aged have, with a few exceptions (Jazdowska 1992, Caldwell *et al.* 1993, HelpAge International 1997) been largely ignored in the literature. Studies of AIDS orphanhood are almost exclusively concerned with the welfare of the orphan rather than of the caregiver (eg Foster *et al.* 1995, Kamali *et al.* 1996, Sengendo and Nambi 1997, Urassa *et al.* 1997, Ntozi 1997b). Concern for the welfare of AIDS orphans has led to an awareness of the need to support their caregivers, who, since at least one of the children's parents has died, are frequently their grandparents (Beer *et al.* 1988, Tout 1989, Tlou 1996). It is notable that, prior to the AIDS epidemic, no support was routinely offered to those grandparents who were caring for non-AIDS orphans, who were co-habiting with a grandchild who helped them in the performance of domestic tasks, or to those whose livelihoods were so precarious that they were unable to care for orphans. It appears that, currently, the concern is for the AIDS orphan rather than for other orphans or for the aged, and while NGOs are currently supporting grandparents in their task of caring for orphans, whether support will continue to be offered to grandparents after the orphans have left their households remains to be seen.

Ainsworth and Over (1994) make the important point that not all AIDS orphans are disadvantaged, and nor are all disadvantaged children AIDS orphans, and that there is a need for effective assessment of their needs to permit targeting of resources to the most needy. Many of the members of the URAA supported association mentioned above were caring for AIDS orphans, and for more orphans than was the case for all but one household in Kikole. Some of these were regularly supplied with food by a charitable organisation, the aim of which was, explicitly, the support of orphans, and not the support of their aged caregivers. The village's location, astride a tarred road some ten kilometres from a major town, may have influenced its choice as a target for this support. Reference to members' wealth rankings revealed that those receiving the support were not the poorest among them, but were those caring for the largest numbers of orphans. Amongst this group were several relatively wealthy aged individuals, who were less in need of support than many of the poorer aged people not caring for orphans.

Empowering interventions directed at aged adults are likely, in the first instance, only to allow the aged person to better support those for whom he or she is responsible, and therefore not to allow the accumulation of additional resources by the aged recipient. Such interventions, though, are likely to prevent or slow the consumption of resources at a relatively early stage of the decline of the aged person's capabilities and assets, and thus prevent or delay the fall into near-destitution that will be the lot of most unsupported aged people. Empowering interventions can postpone the moment at which the aged person will become dependent on others, and provide him or her with a longer period of old age lived in satisfactory circumstances. In the (currently unlikely) event that the state is providing support to the dependent aged, the numbers entitled to this support at any one moment will be reduced if self-sustaining supportive interventions are in existence.

It is encouraging to note some recent indications that the welfare of elders is becoming more widely recognised as a priority development issue. In April 2002 the Second World Congress on Ageing was held in Madrid. This was attended by representatives of nearly 160 countries, and informed by an NGO World Forum, the Valencia Forum (an academic and scientific gathering held immediately prior to the Congress), and documents such as the *World Labour Report 2000* (International Labour Organisation 2000), *Health and Ageing* (World Health Organisation 2001), and the Research Agenda on Ageing for the 21st Century (United Nations 2002a). The Congress adopted an International Plan of Action on Ageing 2002 (United Nations 2002b), and made a statement of political commitment in three priority directions: "older persons and development, advancing health and well-being into old age, and ensuring enabling and supportive environment" (United Nations 2002c). Together, these documents provide the foundations for programmes to improve the well-being of elders in the twenty-first century. However, we must wait to learn whether individual governments are prepared to back their political commitments with financial support.

The work of NGOs is increasingly significant and therefore increasingly valued. HelpAge International, the leading international NGO, is now routinely represented on global peak bodies, and has produced invaluable documents which both detail the problems currently faced by elders (Randel and German 1999, HelpAge International 2002), and present further refinements of strategies to ensure their participation in policy development (Heslop 2002b).

For most of us living in developed countries, the aged in the developing world lead distant lives that differ so much from our own that we find it difficult to envisage the extent of their deprivation. Perhaps it is not surprising that, in developed countries there is little understanding of or sympathy for the hardships of the aged in the poorest countries of the world, and that development programmes frequently ignore the needs of the aged. The aged in other cultures rarely figure in appeals for international charitable support, where the appealing eyes and swollen bellies of starving children are found more productive than the lined faces and stooped bodies of the aged.

I have confined my study to a single, small, aged population, amongst whom I have demonstrated extensive variation in experience and circumstance, but a convergence, with advancing age, towards a homogeneity of need. Having consumed all their assets, and in the absence of support from their children, the rich heterogeneity of their lives becomes subsumed in the common experience of hunger and sadness, and death after a period of malnutrition and social isolation. Although the relatively small numbers of aged people in developing countries has to date allowed their difficulties to remain largely unrecognised, the predicted increase in the world's aged population will make it increasingly difficult to ignore them. However without support from the international community there is unlikely to be significant improvement in their circumstances.

Bibliography

Abel-Smith B and Rawal P 1992. Can the poor afford "free" health services? A case study of Tanzania, *Health Policy and Planning*, 7:329-341.

Adamchak DJ 1989. Population aging in sub-Saharan Africa: the effects of development on the elderly, *Population and environment*, 10:162-176.

Adeokun LA and Nalwadda RM 1997. Serial marriages and AIDS in Masaka district, *Health Transition Review*, 7 (suppl):49-66.

Ainsworth M and Over M 1994. AIDS and African development, *World Bank Research Observer*, 9:203-240.

Albert SM and Cattell MG 1994. *Old Age in Global Perspective*, New York: GK Hall.

Anderson S and Kaleeba N 1994. The challenge of AIDS homecare, *World Health*, 47 No. 4:20-22.

Ankrah EM 1991. AIDS and the social side of health, *Social Science and Medicine*, 32:967-980.

Appleton S 1996. Women-headed households and household welfare: an empirical deconstruction for Uganda, *World Development*, 24:1811-1827.

Apt NA 1988. Aging in Africa, In Gort E, *Aging in cross-cultural perspective: Africa and the Americas*, pp. 17-31, New York: Phelps-Stokes Fund.

Apt NA 1992. Trends and prospects in Africa, *Community Development Journal*, 27:130-139.

Apt NA 1996. *Coping with old age in a changing Africa: social change and the elderly Ghanaian*, Aldershot: Avebury.

Arth M 1968. Ideals and behavior. A comment on Ibo respect patterns, *The Gerontologist*, 8:242-244.

Banaji J 1976. Summary of selected parts of Kautsky's The Agrarian Question, *Economy and Society*, 5:2-49.

Barnett T and Blaikie P 1989. AIDS and food production in East and Central Africa, *Food Policy*, 14:2-6.

Barnett T and Blaikie P 1992. *AIDS in Africa*, London: Belhaven Press.

Barnett T, Tumushabe J, Bantebya G, Ssebuliba R, Ngasongwa J, *et al* 1995. The social and economic impact of HIV/AIDS on farming systems and livelihoods in rural Africa: some experience and lessons from Uganda, Tanzania and Zambia, *Journal of International Development*, 7:163-176.

Barth F 1974. On responsibility and humanity: calling a colleague to account, *Current Anthropology*, 15:99-103.

Barton T and Wamai G 1994. *Equity and vulnerability: a situation analysis of women, adolescents and children in Uganda, 1994*, Kampala: The Government of Uganda and the Uganda National Council for Children.

Bayliss-Smith T 1991. Food security and agricultural sustainability in the New Guinea Highlands: vulnerable people, vulnerable places, *IDS Bulletin*, 22:5-11.

Bazarra N 1994. Land Policy and the evolving forms of land tenure in Masindi district, In Mamdani M and Oloka-Onyango L, *Uganda: studies in living conditions, popular movements and constitutionalism*, pp. 17-60, Kampala: JEP.

Beals RL 1975. *The peasant marketing system of Oaxaca, Mexico*, Berkeley: University of California Press.

Beck T 1989. Survival strategies and power among the poorest in a West Bengal village, *IDS Bulletin*, 20:23-32.

Beer C, Rose A and Tout K 1988. AIDS - The grandmother's burden, In Fleming AF, Carballo M, FitzSimmons DW, Bailey MR and Mann J, *The Global Impact of AIDS*, pp. 171-174, New York: Alan R Liss Inc.

Bender DR 1967. A refinement of the concept of household: families, co-residence and domestic functions, *American Anthropologist*, 69:493-504.

Bennett E, Himmavanh V, Salazar F and Williams A 1994. *Dying at home: the experience of four villages in Northeast Thailand*, unpublished Master of Tropical Health thesis, Brisbane: Tropical Health Program, University of Queensland.

Bennett FJ 1963. Custom and child health in Buganda: concepts of disease, *Tropical and Geographical Medicine*, 15:148-157.

Bennett FJ and Mugalula-Mukiibi A 1967. An analysis of people living alone in a rural community in East Africa, *Social Science and Medicine*, 1:97-115.

Bernstein H 1979. African peasantries: a theoretical framework, *Journal of Peasant Studies*, 6:421-443.

Bernstein H, Crow B and Johnson H 1992. *Rural livelihoods: crises and responses*, Oxford: Oxford University Press.

Bibagambah JR 1996. *Marketing of smallholder crops in Uganda*, Kampala: Fountain.

Biggar R 1986. The AIDS problem in Africa, *The Lancet*, 1:79-83.

Bigsten A and Kayizzi-Mugerwa S 1992. Adaption and distress in the urban economy: a study of Kampala households, *World Development*, 20:1423-1441.

Bigsten A and Kayizzi-Mugerwa S 1995. Rural sector responses to economic crisis in Uganda, *Journal of International Development*, 7:181-209.

Bishop J and Scoones I 1994. *The hidden harvest: the role of wild foods in agricultural systems*, London: International Institute for Environment and Development.

Bitawha N, Tumwesigye O, Kabariime P, Tayebwa AKM, Tumwesigye S, *et al* 1997. Herbal treatment of malaria - four case reports from Rukararwe Partnership Workshop for Rural Development, *Tropical Doctor*, Suppl 1:17-19.

Blaikie P and Brookfield H 1987. *Land degradation and society*, London: Methuen.

Bledsoe C and Goubaud MF 1985. The reinterpretation of Western pharmaceuticals among the Mende of Sierra Leone, *Social Science and Medicine*, 21:275-282.

Bohannon P 1963. *Social Anthropology*, London: Holt, Rinehart and Winston.

Bond GC and Vincent J 1991. Living on the edge: changing social structures in the context of AIDS, In Hansen HB and Twaddle M (Eds), *Changing Uganda -The Dilemmas of Structural Adjustment and Revolutionary Change*. pp. 113-129, London: James Currey.

Bongarts J, Reining P, Way P and Conant F 1989. The relationship between male circumcision and HIV infection in African populations, *AIDS*, 3:373-377.

Boserup E 1970. *Women's role in economic development*, New York: St Martin's Press.

Bradley J 1975. Helminths, In Hall SA and Langlands BW, *Uganda atlas of disease distribution*, pp. 19-22, Nairobi: East African Publishing House.

Brett EA 1992. *Providing for the rural poor: institutional decay and transformation in Uganda*, Brighton: Institute of Development Studies.

Brown RE and Wilks NE 1966. Health survey in Ugandan primary schools, *Tropical and Geographical Medicine*, 18:183-220.

Bunker SG 1983. Dependency, inequality and development policy: a case from Bugisu, Uganda, *The British Journal of Sociology*, 34:182-207.

Cabral AJR 1993. AIDS in Africa: can the hospitals cope?, *Health Policy and Planning*, 8:157-160.

Cairncross S and Feachem R 1993. *Environmental health engineering in the tropics*, Chichester: Wiley.

Caldwell JC 1993. Health Transition: the cultural, social and behavioural determinants of health in the third world, *Social Science and Medicine*, 36:125-135.

Caldwell J, Caldwell P, Ankrah EM, Anarfi JK, Agyeman DK, *et al* 1993. African families and AIDS: context, reactions and potential interventions, *Health Transition Review*, 3 (Suppl):1-16.

Caldwell JC and Caldwell P 1994. The neglect of an epidemiological explanation for the distribution of HIV/AIDS in sub-Saharan Africa: exploring the male circumcision hypothesis, *Health Transition Review*, 4(Suppl):23-46.

Campbell ID and Rader AD 1995. HIV counselling in developing countries: the link from individual to community counselling for support and change, *British Journal of Guidance and Counselling*, 23:33-43.

Carballo M and Carael M 1988. Impact of AIDS on social organisation, In Fleming AF, Carballo M, FitzSimmons DW, Bailey MR and Mann J, *The Global Impact of AIDS*, pp. 81-93, New York: Alan R Liss Inc.

Cardoso FH 1972. Dependence and development in Latin America, *New Left Review*, July-August 1972:83-95.

Carpenter L 1998. *Incidence and prevalence of HIV infection among adults in Kyamulibwa, 1990-1997.* (Pers. comm.).

Carswell JW 1988. Impact of AIDS in the developing world, *British Medical Bulletin*, 44:183-202.

Cattell MG 1989. Knowledge and social change in Samia, Western Kenya, *Journal of Cross-Cultural Gerontology*, 4:225-244.

Chambers R 1988. Sustainable rural livelihoods: a key strategy for people, environment and development, In Conroy C and Litvinoff M, *The greening of AID: sustainable livelihoods in practice*, pp. 1-17, London: Earthscan.

Chambers R 1994. The origins and practice of participatory rural appraisal, *World Development*, 22:953-969.

Chambers R 1995. *Poverty and livelihoods: whose reality counts?*, Brighton: Institute of Development Studies.

Chambers R 1997. *Whose reality counts? Putting the first last*, London: Intermediate Technology Publications.

Chambers R and Conway GR 1992. *Sustainable rural livelihoods: practical concepts for the 21st century*, Brighton: Institute of Development Studies.

Chambers R, Longhurst R and Pacey A 1981. *Seasonal dimensions to rural poverty*, London: Frances Pinter.

Chandra RK 1983. Nutrition, immunity and infection: present knowledge and future directions, *The Lancet*, 1:688-691.

Chin J 1991. The epidemiology and projected mortality of AIDS, In Feachem RG and Jamison DT, *Disease and mortality in Sub-Sarahan Africa*, pp. 203-213, Oxford: Oxford University Press.

Chirimuuta RC and Chirimuuta RJ 1989. *AIDS, Africa and racism*, London: Free Association Books.

Chrisman N 1977. The health seeking process, *Culture, Medicine and Psychiatry*, 1:351-372.

Clumeck N, Mascart-Lemone F, de Mauberge J and Marcelis L 1983. Acquired immune deficiency syndrome in black Africans, *The Lancet*, 1:642.

Cohen D 1998. *The HIV epidemic and sustainable human development*, New York: UNDP.

Colson AC 1972. The differential use of medical resources in developing countries, *Journal of Health and Social Behavior*, 12:226-237.

Colson E and Scudder T 1981. Old age in Gwembe district, Zambia, In Amoss PT and Harrell S, *Other Ways of Growing Old*, pp. 125-154, Stanford: Stanford University Press.

Corbett J 1988. Famine and household coping strategies, *World Development*, 16:1099-1112.

Corbett J 1989. Poverty and sickness: the high cost of ill-health, *IDS Bulletin*, 20(2):58-62.

Cowgill DO 1986. *Aging around the world*, Belmont: Wadsworth.

Cowgill DO and Holmes LD 1972. Summary and conclusions: the theory in review, In Cowgill DO and Holmes LD, *Aging and Modernisation*, pp. 305-323, New York: Meredith Corporation.

Cronbach LJ 1951. Coefficient alpha and the internal structure of tests, *Psychometrika*, 16:297-334.

Crystal S 1989. Persons with AIDS and older people: common long-term care concerns, In Riley M, Ory M and Zablotsky D, *AIDS in an Aging Society*, pp. 147-166, New York: Springer.

Danziger R 1994. The social impact of HIV/AIDS in developing countries, *Social Science and Medicine*, 39:905-917.

Davachi F, Baudoux P, Ndoko K, N'Galy B and Mann J 1988. The economic impact on families of children with AIDS in Kinshasa, Zaire, In Fleming AF, Carballo M, FitzSimmons DW, Bailey MR and Mann J, *The Global Impact of AIDS*, pp. 167-169, New York: Alan R Liss Inc.

Davey KJ 1974. *Taxing a peasant society: the example of graduated taxes in East Africa*, London: Charles Knight.

Davies S 1993. Are coping strategies a cop out?, *IDS Bulletin*, 24:60-72.

Davies S 1996. *Adaptable Livelihoods: coping with food insecurity in the Malian Sahel*, London: MacMillan.

Dawson S, Manderson L and Tallo VL 1993. *A Manual for Focus Groups*, Boston: International Nutrition Foundation for Developing Countries.

De Cock KM, Lucas SB, Lucas S, Agness J, Kadio A, *et al* 1993. Clinical research, prophylaxis, therapy, and care for HIV disease in Africa, *American Journal of Public Health*, 83:1385-1389.

de Garine I and Harrison GA 1988. *Coping with uncertainty in food supply*, Oxford: Clarendon Press.

Devereux S and Eele G 1991. *The Social and Economic Impact of AIDS in East and Central Africa*, Oxford: Food Studies Group.

Devereux SD 2001. *Social pensions in Namibia and South Africa.* IDS Working Paper 370. Brighton: Institute of Development Studies.

Directorate of Overseas Surveys 1961. *Rainfall probability, population and main drainage basin*, Map 299D. Scale 1:4,000,000. London: Directorate of Overseas Surveys.

Dodge CP 1987. Rehabilitation or redefinition of health services, In Wiebe PD and Dodge CP, *Beyond crisis: development issues in Uganda*, pp. 101-112, Kampala: Makerere Institute of Social Research and the African Studies Association.

Dodge CP and Wiebe PD 1985. *Crisis in Uganda: the breakdown of health services*, Oxford: Pergamon.

Dorjahn VR 1989. Where do the old folks live? The residence of the elderly among the Temne of Sierra Leone, *Journal of Cross-Cultural Gerontology*, 4:257-278.

Dos Santos T 1970. The structure of dependence, *American Economic Review*, 60:231-236.

Drewnowski J and Scott W 1966. *The level of living index*, Geneva: United Nations Research Institute for Social Development.

Dreze J and Srinivasan PV 1995. *Widowhood and poverty in rural India: some inferences from household survey data*, London: Suntory-Toyota International Centre for Economics and Related Disciplines.

du Boulay J and Williams R 1984. Collecting Life Histories, In Ellen RF, *Ethnographic Research: a guide to general conduct*, pp. 247-257, London: Academic Press.

du Guerny J 1997. *The rural elderly and the ageing of rural populations* Paper presented at the World Congress of Gerontology/Inter Agency meeting for the International Year of Older Persons, Adelaide, Australia, 19-22 August 1997.

Dunn A 1992. *The Social consequences of HIV/AIDS in Uganda*, London: Save the Children.

du Toit BM 1990. People on the move: rural urban migration with special reference to the third world: theoretical and empirical perspectives, *Human Organisation*, 49:305-319.

Ellis F 1988. *Peasant economies: farm households and agrarian development*, Cambridge: Cambridge University Press.

Escobar A 1995. *Encountering development: the making and unmaking of the third world*, Princeton: Princeton University Press.

Evans-Pritchard EE 1937. *Witchcraft, oracles and magic among the Azande*, Oxford: Clarendon Press.

Eveleth PB and Tanner JM 1990. *Worldwide variation in human growth*, Cambridge: Cambridge University Press.

Fallers LA 1964a. Social stratification in traditional Buganda, In Fallers LA, *The King's Men*, pp. 64-116, London: Oxford University Press.

Fallers LA 1964b. Social mobility, traditional and modern, In Fallers LA, *The King's Men: leadership and status in Buganda on the eve of independence*, pp. 158-210, Oxford: Oxford University Press.

Fallers LA 1964c. The modernisation of social stratification, In Fallers LA, *The King's Men*, pp. 117-157, London: Oxford University Press.

Fallers LA 1965. *Bantu Bureaucracy*, Chicago: University of Chicago Press.

Feachem RG, Jamison DT and Bos ER 1991. Changing patterns of disease and mortality in sub-Saharan Africa, In Feachem RG and Jamison DT, *Disease and mortality in sub-Saharan Africa*, pp. 3-28, Oxford: Oxford University Press.

Fleischman J 1995. Living HIV-positively, *Africa Report*, 40:56-59.

Fleuret A 1986. Survey of nutrition in East Africa, In Quandt S and Ritenough C, *Training Manual in Nutritional Anthropology*, pp. 90-97, Washington DC: American Anthropological Society.

Folta JR and Deck ES 1987. Elderly black widows in rural Zimbabwe, *Journal of Cross-Cultural Gerontology*, 2:321-342.

Foner N 1984. *Ages In Conflict: A cross-cultural perspective on inequality between old and young*, New York: Columbia University Press.

Food and Agriculture Organisation 1987. Women in African food production and food security, In Gittinger JP, *Food Policy*, pp. 133-140, Washington DC: World Bank.

Fortt JM and Hougham DA 1973. Environment, population and economic history, In Richards AI, *Subsistence to commercial farming in present day Buganda*, pp. 17-46, Cambridge: Cambridge University Press.

Foster G, Shakespeare R, Chinemana F, Jackson H, Gregson S, *et al* 1995. Orphan prevalence and extended family care in a peri-urban community in Zimbabwe, *AIDS Care*, 7:3-17.

Frankenberger TR 1993. Promoting sustainable livelihoods, *Arid Lands Newsletter*, 34:30-42.

Frankenberger TR and Goldstein DM 1990. Food security, coping strategies and environmental degradation, *Arid Lands Newsletter*, 20:21-27.

Friedan B 1994. *The fountain of age*, London: Vintage.

Gaisie S 1989. Culture and health in sub-Saharan Africa, In Caldwell J, Findley S, Caldwell P, Santow G, Cosford W, *et al*, *What do we know about the health transition?*, pp. 609-627, Canberra: Australian National University Health Transition Centre.

Gershenberg I 1972. The distribution of medical services in Uganda, *Social Science and Medicine*, 6:353-372.

Gibson MJ 1992. Public health and social policy, In Kendig HL, Hashimoto A and Coppard L, *Family support to the elderly: the international experience*, pp. 88-114, Oxford: Oxford University Press.

Gibson RW, Legg JP and Otim-Nape GW 1996. Unusually severe symptoms are a characteristic of the current epidemic of mosaic virus disease of cassava in Uganda, *Annals of Applied Biology*, 128:479-490.

Giesler G 1993. Silences speak louder than claims: gender, household and agricultural development in Southern Africa, *World Development*, 21:1965-1980.

Gill GJ 1991. *Seasonality and agriculture in the developing world*, Cambridge: Cambridge University Press.

Gittinger J 1990. *Household food security and the role of women*, Washington DC: World Bank.

Glascock AP 1986. Resource control among the older males in Southern Somalia, *Journal of Cross-Cultural Gerontology*, 1:51-72.

Goldin CS 1994. Stigmatization and AIDS: critical issues in public health, *Social Science and Medicine*, 39:1359-1366.

Goldman N, Korenman S and Weinstein R 1995. Marital status and health among the elderly, *Social Science and Medicine*, 40:1717-1730.

Good CM 1987. *Ethnomedical Systems in Africa*, New York: The Guilford Press.

Goodgame RW 1990. AIDS in Uganda - clinical and social features, *New England Journal of Medicine*, 323:383-389.

Gore MS 1990. Social factors affecting the health of the elderly, In Kane R, Evans JG and MacFadyen D, *Improving the Health of Older People: a World View*, pp. 107-123, New York: Oxford University Press.

Gorman M 1995. Older people and development: the last minority?, *Development in Practice*, 5:117-127.

Gould WTS 1988. Rural-urban interaction and rural transformation in tropical Africa, In Rimmer D, *Rural transformation in tropical Africa*, pp. 77-97, London: Belhaven.

Grandin BE 1986. *Wealth Ranking in Smallholder Communities: a field manual*, London: Intermediate Technology Group.

Green RH 1981. *Magendo in the political economy of Uganda: pathology, parallel system or dominant sub-mode of production?* Brighton: IDS.

Griffiths JF 1962. The climate of East Africa, In Russell EW, *The natural resources of East Africa*, pp. 77-87, Nairobi: East African Literature Bureau.

Gubrium JF 1973. *The myth of the golden years*, Springfield: Thomas.

Guijit I, Fuglesang A and Kisadha T 1994. *It is the young trees that make a thick forest*, Kampala and London: Redd Barna and International Institute for Environment and Development.

Guyer J 1981. Household and community in African studies, *African Studies Review*, 24:87-137.

Hampton J 1991. *Living positively with AIDS*, London: ActionAid.

Harriss J 1987. Capitalism and peasant production: the green revolution in India, In Shanin T, *Peasants and peasant societies*, pp. 227-246, Oxford: Blackwell.

Harwood A 1971. The hot-cold theory of diseases, *Journal of the American Medical Association*, 216:1153-1158.

Hashimoto A and Kendig HL 1992. Aging in international perspective, In Kendig HL, Hashimoto A and Coppard LC, *Family support to the elderly: the international experience*, pp. 3-14, Oxford: Oxford University Press.

Haslwimmer M 1994. Is HIV/AIDS a threat to livestock production? The example of Rakai, Uganda, *World Animal Review*, 80-81:92-106.

Helman CG 1994. *Culture, health and illness*, Oxford: Butterworth-Heinemann.

HelpAge International 1993. *Some methods for interactive learning with elders (SMILE)*, Nairobi: HelpAge Kenya.

HelpAge International 1995a. *Assessment into the needs of older people in Elim, Northern Transvaal Province, Republic of South Africa*, London: HelpAge International.

HelpAge International 1995b. *Needs assessment in Changara district, for development of resettlement programme in Tete*, London: HelpAge International.

HelpAge International 1997. *Intergenerational involvement in HIV/AIDS prevention and care. New Delhi; 11-14 March 1997*, London: HelpAge International.

HelpAge International 2002. *The state of the world's older people 2002*, London: HelpAge International.

Hendricks J and Hendricks CD 1977. The age old question of old age: was it really so much better back when?, *Journal of Aging and Human Development*, 8:139-154.

Heslop A 2002a. *Livelihoods and coping strategies of older people: the role of non-contributory pensions in the developing world*, London: HelpAge International.

Heslop A 2002b. *Participatory research with older people: a sourcebook*, London: HelpAge International.

Hill P 1972. *Rural Hausa: a village and its setting*, Cambridge: Cambridge University Press.

Hobley CW 1910. *Bantu beliefs and magic*, London: Frank Cass.

Hooper E 1987. AIDS in Uganda, *African Affairs*, 86:467-477.

Hougham DA and Sturrock F 1973. The farms – present day organisation, In Richards AI, Sturrock F and Fortt JM, *Subsistence to commercial farming in present day Buganda*, pp. 150-178, Cambridge: Cambridge University Press.

Hunt CW 1989. Migrant labour and sexually transmitted disease: AIDS in Africa, *Journal of Health and Social Behavior*, 30:353-373.

Hunter SS 1990. Orphans as a window on the AIDS epidemic in Africa: initial results and implications of a study in Uganda, *Social Science and Medicine*, 31:681-690.

Hunter SS, Bulirwa E and Kisseka E 1993. AIDS and agricultural production: reports of a land utilization survey, Masaka and Rakai Districts of Uganda, *Land Use Policy*, 10:241-258.

Iliffe J 1987. *The African Poor*, Cambridge: Cambridge University Press.

Ingstad B, Bruun F, Sandberg E and Tlou S 1992. Care for the elderly, care by the elderly: the role of elderly women in Tswana society, *Journal of Cross-Cultural Gerontology*, 7:379-398.

International Labour Organisation 2000. *World Labour Report 2000: income security and social protection in a changing world*, Geneva: International Labour Organisation.

Jaenson C, Harmsworth J, Kabwegyere T and Muzaale P 1984. *The Uganda social and institutional profile*, Kampala: USAID.

Jager H, Jersild C and Emmanuel JC 1991. Safe blood transfusions in Africa, *AIDS*, 1(Suppl 1):S163-S168.

Jamal V 1991. The agrarian context of the Ugandan crisis, In Hansen HB and Twaddle M (Eds), *Changing Uganda -The Dilemmas of Structural Adjustment and Revolutionary Change.* pp. 78-97, London: James Currey.

Jazdowska N 1992. *Elderly women caring for orphans and people with AIDS*, Harare: HelpAge Zimbabwe.

Jodha NS 1988. Poverty debate in India: a minority view, *Economic and Political Weekly*, Special Issue November 1988:2421-2428.

Jorgensen D 1989. *Participant Observation: A Methodology for Human Studies*, Newbury Park: Sage Publications Inc.

Kaduru G, Mwesigye E and Nambi A 1990. *A preliminary report on a needs assessment study of Rakai and Masaka districts with particular reference to the socio-economic impact of the AIDS epidemic*, Kampala: Lutheran World Federation.

Kaggwa A 1951. *Engero za Baganda*, London: Sheldon Press.

Kajubi S 1985. Integration and national development from the viewpoint of education in Uganda, In Dodge CP and Wiebe PD, *Crisis in Uganda: the breakdown of the health services*, pp. 15-24, Oxford: Pergamon.

Kakande ML, Bennett FJ and Rawji F 1972. Selected aspects of the health of old people in rural Buganda, *East African Medical Journal*, 49:970-982.

Kamali A, Seeley JA, Nunn AJ, Kengeya-Kayondo J-F, Ruberantwari A, and Mulder DW 1996. The orphan problem: experience of a sub-Saharan Africa rural population in the AIDS epidemic. AIDS Care 8:509-515

Kamali A, Wagner H, Nakiyingi J, Sabiiti I, Kengeya-Kayondo JF, *et al* 1996. Verbal autopsy as a tool for diagnosing HIV-related adult deaths in rural Uganda, *International Journal of Epidemiology*, 25:679-684.

Kane RL, Ouslander JG and Abrass IB 1994. *Essentials of clinical geriatrics*, New York: McGraw-Hill.

Kasfir N 1976. *The shrinking political arena: participation and ethnicity in African politics, with a case study of Uganda*, Berkeley: University of California Press.

Kaufmann T 1993. *A crisis of silence: HIV, AIDS and older people*, London: Age Concern.

Kaufmann T 1995. *HIV and AIDS and older people*, London: Age Concern England.

Kayongo-Male D and Onyango P 1984. *The Sociology of the African Family*, London: Longman.

Kendig H 1987. Roles of the aged, families and communities in the context of an aging society, In United Nations, *Population Aging: review of emerging issues*, pp. 75-83, Bangkok: United Nations.

Kengeya-Kayondo JF, Seeley JA, Kajura-Bajenja E, Kabunga E, Mubiru E, *et al* 1994. Recognition, treatment seeking behaviour and perception of cause of malaria among rural women in Uganda, *Acta Tropica*, 58:267-273.

Kengeya-Kayondo J-F, Malamba SS, Nunn AJ, Seeley JA, Ssali A, *et al* 1995. Human immunodeficiency virus (HIV-1) seropositivity among children in a rural population of South-west Uganda: probable routes of exposure, *Annals of Tropical Paediatrics*, 15:115-120.

Kengeya-Kayondo J-F, Kamali A, Nunn AJ, Ruberantwari A, Wagner H, *et al* 1996. Incidence of HIV-1 infection in adults and socio-demographic characteristics of sero-converters in a rural population in Uganda, *International Journal of Epidemiology*, 25:1077-1082.

Keyfitz N and Flieger W 1990. *World Population Growth and Aging*, Chicago: University of Chicago Press.

Khasiani SA 1987. The role of the family in meeting the social and economic needs of the aging population in Kenya, *Genus*, 43:103-118.

Khasiani SA 1991. *The integration of ageing and elderly women into development in Eastern Africa*, New York: United Nations.

Kilbride PL and Kilbride JC 1990. *Changing family life in East Africa*, University Park: Pennsylvania State University Press.

Kinsella K 1997. The demography of an aging world, In Sokolovsky J, *The cultural context of aging: worldwide perspectives*, pp. 17-32, Westport: Bergin and Garvey.

Kinsella K and Taeuber C 1993. *An Aging World II*, Washington DC: United States Bureau of the Census.

Kironde E 1985. A village perspective of health services in Buganda, In Dodge C and Wiebe P, *Crisis in Uganda: The Breakdown of Health Services*, pp. 65-68, Oxford: Pergamon Press.

Kiwanuka MSMS 1971. *A history of Buganda from the foundation of the Kingdom to 1900*, London: Longman.

Kjellstrom T and Rosenstock L 1990. The role of environmental and occupational hazards in the adult health transition, *World Health Statistics Quarterly*, 43:188-196.

Kleinman A 1980. *Patients and healers in the Context of Culture*, Berkeley: University of California Press.

Kloos H 1994. The poorer third world: health and health care in areas that have yet to experience substantial development, In Phillips D and Verhasselt Y, *Health and Development*, pp. 199-215, London: Routledge.

Kolsrud T, Amanyire M, Landbo G, Jareg P, Byangire, *et al* 1989. *Social consequences of AIDS in Uganda*, Oslo: Redd Barna.

Kreniske J 1991. *HIV transmission and economic development*, New York: HIV Centre for Clinical and Behavioral Studies, Columbia University.

Lado C 1992. Female labour participation in agricultural production and the implications for nutrition and health in rural Africa, *Social Science and Medicine*, 34:789-807.

Langlands BW 1965. Maize in Uganda, *Uganda Journal*, 29:215-221.

Laslett P 1965. *The world we have lost*, London: Methuen.

Last JM 1992. Housing and health, In Last JM and Wallace RB, *Public health and preventive medicine*, pp. 671-675, London: Prentice-Hall.

Linsk NL 1994. HIV and the elderly, *Journal of Contemporary Human Services*, 75:362-372.

Lipton M 1988. *The poor and the poorest: some interim findings*, Washington DC: World Bank.

Longhurst R 1983. Agricultural production and food consumption: some neglected linkages, *Food and Nutrition*, 9(2):2-.

Longhurst R 1988. Cash crops and food security, *IDS Bulletin*, 19:28-36.

Lubega D 2001. *Financing of education in UPE schools*, Kampala: Centre for Basic Research.

Macrae J, Zwi AB and Gilson L 1996. A triple burden for health sector reform: "post-conflict" rehabilitation in Uganda, *Social Science and Medicine*, 42:1095-1108.

Mair L 1933. Baganda land tenure, *Africa*, 6:187-205.

Mair L 1934. *An African People in the Twentieth Century*, London: George Routledge and Sons Ltd.

Mair L 1940. *Native marriage in Buganda*, London: Oxford University Press.

Mair L 1957. The contribution of social anthropology to the study of changes in African land tenure, In Mair LP, *Studies in Applied Anthropology*, pp. 53-62, London: Athlone Press.

Mair L 1965. *An Introduction to Social Anthropology*, Oxford: Clarendon Press.

Malamba SS, Wagner H, Maude G, Okongo M, Nunn AJ, *et al* 1994. Risk factors for HIV-1 infection in adults in a rural Ugandan community: a case-control study, *AIDS*, 8:253-257.

Mamdani M 1976. *Politics and Class Formation in Uganda*, New York: Monthly Review Press.

Mamdani M 1987. Extreme but not exceptional: towards an analysis of the agrarian question in Uganda, *Journal of Peasant Studies*, 14:191-225.

Mamdani M 1996. Analysing the agrarian question: the case of a Buganda village, In Mamdani M, *Uganda: studies in labour*, pp. 325-352, Dakar: Council for the development of social science research in Africa (CODESRIA.

Mann JM, Chin J, Piot P and Quinn T 1988. The international epidemiology of AIDS, *Scientific American*, 259:82-89.

Maquet JJ 1953. Kinship groups in old Rwanda, *Africa*, 23:25-28.

Mauss M 1970. *The Gift*, London: Cohen and West.

Maxwell RJ and Silverman P 1970. Information and esteem: cultural considerations in the treatment of the aged, *International Journal of Aging and Human Development*, 1:361-392.

Mcgrath JW, Rwabukwali CB, Schumann DA, Pearson-Marks J, Nakayiwa S, *et al* 1993. Anthropology and AIDS: the cultural context of sexual risk behaviour among urban Baganda women in Kampala, Uganda, *Social Science and Medicine*, 36:429-439.

Ministry of Health 1991, *Health Information Quarterly*, 8(3) September. Entebbe: Ministry of Health.

Ministry of Local Government 1960. *Memorandum for the guidance of African authorities in the development of their direct personal taxation systems*, Entebbe: Ministry of Local Government.

Moore SF 1978. Old age in a life-term social arena: some Chagga of Kilimanjaro in 1974, In Myerhoff BG and Simic A, *Life's Career - Aging*, pp. 23-75, Beverley Hills: Sage.

Morgan D, Malamba SS, Maude GH, Okongo MJ, Wagner H, *et al* 1997. An HIV-1 natural history cohort and survival times in rural Uganda, *AIDS*, 11:633-640.

Morgan D, Mahe C, Mayanja B, Okongo JM, Lubega R, and Whitworth JAG, 2002. HIV-1 infection in rural Africa: is there a median time to AIDS and survival compared to that in industrialized countries? *AIDS* 16:597-603.

Muhereza F 1994. Land tenure and peasant adaptations: some reflections on agricultural production in Luwero district, In Mamdani M and Oloka-Onyango L, *Uganda: studies in living conditions, popular movements and constitutionalism*, pp. 61-96, Kampala: JEP.

Mukasa S 1971 [first publ. c1925]. Autobiography of a Muganda chief, In Low DA, *The mind of Buganda: documents of the modern history of an African kingdom*, pp. 57-61, London: Heinemann.

Mulder DW, Nunn AJ, Kamali A, Nakiyingi J, Wagner H, *et al* 1994. Two-year HIV-1 associated mortality in a Ugandan rural population, *The Lancet*, 343:1021-1023.

Mulder DW, Nunn A, Kamali A and Kengeya-Kayondo JF 1996. Post-natal incidence of HIV-1 infection among children in rural Ugandan population: no evidence for transmission other than mother to child, *Tropical Medicine and International Health*, 1:81-85.

Muller O and Abbas N 1990. The impact of AIDS mortality on children's education in Kampala (Uganda), *AIDS Care*, 2:77-80.

Murphy JD 1972. *Luganda-English Dictionary*, Washington DC: Consortium.

Muyinda H, Seeley J, Pickering H and Barton T 1997. Social aspects of AIDS-related stigma in rural Uganda, *Health and Place*, 3:143-147.

Myers GC 1992. Demographic aging and family support for older persons, In Kendig HL, Hashimoto A and Coppard LC, *Family support to the elderly: the international experience*, pp. 31-68, Oxford: Oxford University Press.

Nabaitu J, Bachengana C and Seeley J 1994. Marital instability in a rural population in South-west Uganda: implications for the spread of HIV-1 infection, *Africa*, 64:243-251.

Nabarro D and McConnell C 1989. The impact of AIDS on socioeconomic development, *AIDS*, 3 (suppl 1):S265-S272.

Nahemow N 1979. Residence, kinship and social isolation among the aged Baganda, *Journal of Marriage and the Family*, :171-183.

Nahemow N 1987. Grandparenthood among the Baganda: Role option in old age, In Sokolovsky J, *Growing Old in Different Societies: cross-cultural perspectives*, pp. 104-115, Acton, Mass: Copley.

Nahemow N and Adams BN 1974. Old age among the Baganda: continuity and change, In Gubrium JF, *Late Life: communities and environmental policy*, pp. 147-166, Springfield: Thomas.

Nayar US 1996. The situation of aging: the chip and the block, In United Nations, *Added years of life in Asia*, pp. 59-81, Bangkok: United Nations.

Ndeti K 1987. Stress and aging in Kenya, In Levi L, *Society, Stress and Disease*, pp. 101-104, Oxford: Oxford University Press.

Neysmith SM 1990. Dependency among third world elderly: a need for new direction in the nineties, *International Journal of Health Services*, 20:681-690.

Neysmith SM and Edwardh J 1984. Economic dependency in the 1980s: its impact on third world elderly, *Ageing and Society*, 4:21-44.

Norton A, Aryeetey E B, Korboe D and Dogbe DKT 1995. *Poverty assessment in Ghana using qualitative and participatory research methods*, Washington DC: World Bank.

Nsibambi AR 1996. Land tenure relations in Uganda 1900-1995, *Uganda Journal*, 43:12-33.

Nsimbi MB 1956. Village life and customs in Buganda, *Uganda Journal*, 20:27-36.

Nsimbi MB 1964. The clan system in Buganda, *Uganda Journal*, 28:25-30.

Ntozi JPM 1997a. AIDS morbidity and the role of the family in patient care in Uganda, *Health Transition Review*, 7 (suppl):1-22.

Ntozi JPM 1997b. Effect of AIDS on children: the problem of orphans in Uganda, *Health Transition Review*, 7 (suppl):23-40.

Ntozi JPM and Mukiza-Gapere J 1995. Care for AIDS orphans in Uganda: findings from focus group discussions, *Health Transition Review*, 5 (suppl):245-252.

Nugent JB 1990. Old age security and the defense of social norms, *Journal of Cross-Cultural Gerontology*, 5:243-254.

Nunn AJ, Kengeya-Kayondo JF, Malamba S, Seeley JA and Mulder DW 1994. Risk factors for HIV-1 infection in adults in a rural Ugandan community: a population study, *AIDS*, 8:81-86.

Nunn AJ, Wagner HU, Okongo JM, Malamba SS and Mulder DW 1996. HIV-1 infection in a Ugandan town on the trans-African highway: prevalence and risk factors, *International Journal of STD and AIDS*, 7:123-130.

Nunn AJ, Mulder DW, Kamali A, Ruberantwari A, Kengeya-Kayondo JF, *et al* 1997. Mortality associated with HIV-1 infection over five years in a rural Ugandan population: cohort study, *British Medical Journal*, 315(7111):767-771.

Nydegger CN 1983. Family ties of the aged in cross-cultural perspective, *The Gerontologist*, 23:26-32.

Obbo C 1986. Some East African widows, In Potash B, *Widows in African Societies*, pp. 84-106, Stanford: Stanford University Press.

Obbo C 1991. Women, children and a "living wage", In Hansen HB and Twaddle M (Eds), *Changing Uganda - The Dilemmas of Structural Adjustment and Revolutionary Change*, pp. 98-112, London: James Currey.

Obbo C 1993. HIV transmission through social and geographical networks in Uganda, *Social Science and Medicine*, 36:949-955.

Offenstadt G, Pinta P, Hericord P, Jagueux M, Jean F, *et al* 1983. Multiple opportunistic infections due to AIDS in a previously healthy black woman from Zaire, *New England Journal of Medicine*, 308:775.

Ogawa N 1992. Resources for the elderly in economic development, In Kendig HL, Hashimoto A and Coppard LC, *Family support to the elderly: the international experience*, pp. 69-87, Oxford: Oxford University Press.

Okello DO, Lubanga R, Guwatudde D and Sebina-Zziwa A 1998. The challenge to restoring basic health care in Uganda, *Social Science and Medicine*, 46:13-21.

Okojie FA 1988. Aging in sub-Saharan Africa: towards a redefinition of needs research and policy directions, *Journal of Cross-Cultural Gerontology*, 3:3-19.

Okuonzi SA and Macrae J 1995. Whose policy is it anyway? International and national influences on health policy development in Uganda, *Health Policy and Planning*, 10:122-132.

Olshansky SJ and Ault AB 1986. The fourth stage of the epidemiologic transition: the age of delayed degenerative diseases, *Millbank Quarterly*, 64:355-391.

Omran AR 1971. The epidemiologic transition, *Millbank Fund Memorial Quarterly*, 49:509-538.

Ouma SOA 1995. The role of social protection in the socioeconomic development of Uganda, *Journal of Social Development in Africa*, 10:5-12.

Over M, Ellis RP, Huber JH and Solon O 1992. The consequences of adult ill-health, In Feachem R et al, *The health of adults in the developing world*, pp. 161-207, New York: Oxford University Press/World bank.

Owen M 1996. *A world of widows*, London: Zed Books.

Palma G 1978. Dependency: a formal theory of underdevelopment or a methodology for the analysis of concrete situations of underdevelopment?, *World Development*, 6:881-924.

Palmore E and Manton K 1974. Modernisation and the status of the aged: international correlations, *Journal of Gerontology*, 29:205-210.

PANOS 1992. *The Hidden Cost of AIDS*, London: The PANOS Institute.

Parkin DM, Pisani P and Ferlay J 1993. Estimates of the worldwide frequency of eighteen major cancers in 1985, *International Journal of Cancer*, 54:595-606.

Payne K 1995. *Older women in development*, London: HelpAge International.

Phillips D 1990. *Health and health care in the third world*, Harlow: Longman.

Pickering H, Okongo M, Nnalusiba B, Bwanika K and Whitworth J 1997. Sexual networks in Uganda: casual and commercial sex in a trading town, *AIDS Care*, 9:199-207.

Pinstrup-Anderson P 1983. Export crop production and malnutrition, *Food and Nutrition*, 9(2):6-14.

Piot P, Goemann J and Laga M 1994. The epidemiology of HIV and AIDS in Africa, In Essex M, Mboup S, Kanki PJ and Kalengayi MR, *AIDS in Africa*, pp. 157-171, New York: Raven press.

Pitayanon S, Rerks-ngarm S, Kongsin S and Janjareon S 1994. *Study on the impact of HIV/AIDS mortality on households in Thailand*, Bangkok: Chulalongkorn University.

Piwoz EG and Viteri FE 1985. Studying health and nutrition behaviour by examining household decision-making, intra-household resource distribution and the role of women in these processes, *Food and Nutrition Bulletin*, 7:1-31.

Planning Dept 1989. *Agricultural sector survey, 1986-87*, Entebbe: Ministry of Animal Industry and Fisheries.

Potash B 1986. Widows in Africa: an introduction, In Potash B, *Widows in African Societies*, pp. 1-43, Stanford: Stanford University Press.

Pratt B and Boyden J 1985. *The field director's handbook: an Oxfam manual for development workers*, Oxford: Oxford University Press.

Preble EA 1990. Impact of HIV/AIDS on African children, *Social Science and Medicine*, 31:671-680.

Prudencio YC and Al-Hassan R 1994. The food security stabilization roles of cassava in Africa, *Food Policy*, 19:57-64.

Radcliffe-Brown AR 1940. Preface, In Fortes M and Evans-Pritchard EE, *African political systems*, p. 302, London: Oxford University Press.

Radcliffe-Brown AR 1950. Introduction, In Radcliffe-Brown AR and Forde D, *African systems of kinship and marriage*, pp. 1-85, London: Oxford University Press.

Rahman MO 1997. The effect of spouses on the mortality of older people in rural Bangladesh, *Health Transition Review*, 7:1-12.

Randell J and German T 1999. *The Ageing and Development Report: poverty, independence and the world's older people*, London HelpAge International.

Ray B 1980. The story of Kintu: myth, death, and ontology in Buganda, In Karp I and Bird CS, *Explorations in African Systems of Thought*, pp. 60-79, Bloomington: Indiana University Press.

Ray B 1991. *Myth, Ritual and Kingship in Buganda*, New York: Oxford University Press.

Richards AI 1964. Authority patterns in traditional Buganda, In Fallers LA, *The King's Men*, pp. 256-293, London: Oxford University Press.

Richards AI 1966. *The Changing Structure of a Ganda Village*, Nairobi: East Africa Publishing House.

Richards AI 1973a. *Economic development and tribal change*, Nairobi: Oxford University Press.

Richards AI 1973b. Authority patterns in traditional Buganda, In Fallers LA, *The King's Men*, pp. 256-293, London: Oxford University Press.

Richards AI, Sturrock F and Fortt JM 1973 (Eds). *Subsistence to commercial farming in present-day Buganda*. Cambridge: Cambridge University Press.

Robertson AF 1978. *Community of Strangers: a journal of discovery in Uganda*, London: Scolar Press.

Robinson-Pant A 1996. PRA: a new literacy?, *Journal of International Development*, 8:531-551.

Roscoe J 1911. *The Baganda*, London: Frank Cass.

Rosenmayr L 1988. More than wisdom: a field study of the old in an African village, *Journal of Cross-Cultural Gerontology*, 3:21-40.

Rowntree BS 1941. *Poverty and Progress: a second social survey of York*, London: Longman.

Russell M 1993. Are households universal? on misunderstanding domestic groups in Swaziland, *Development and Change*, 24:755-785.

Rutishauser IHE 1963. Custom and child health in Buganda: food and nutrition, *Tropical and Geographical Medicine*, 15:138-147.

Rwezaura BA 1989. Changing community obligations to the elderly in contemporary Africa, In Eekelaar J and Pearl D, *An Aging World: Dilemmas and Challenges for Law and Social Policy*, pp. 113-131, Oxford: Clarendon Press.

Sahn DE 1989. A conceptual framework for examining the seasonal aspects of household food security, In Sahn DE, *Seasonal variability in third world agriculture*, pp. 3-16, Baltimore: John Hopkins University Press.

Sangree WH 1987. The childless elderly in Tiriki, Kenya, and Irigwe, Nigeria: a comparative analysis of the relationship between beliefs about childlessness and the social status of the childless elderly, *Journal of Cross-Cultural Gerontology*, 2:201-223.

Sangree WH 1997. Pronatalism and the elderly in Tiriki, Kenya, In Weisner TS, Bradley C and Kilbride PL, *African families and the crisis of social change*, pp. 184-207, Westport: Bergin and Garvey.

Sauerborn R, Ibrango I, Nougtara A, Borchert M, Hien M, *et al* 1995. The economic costs of illness for rural households in Burkina Faso, *Tropical Medicine and Parasitology*, 46:54-60.

Scheyer S and Dunlop D 1985. Health services and development in Uganda, In Dodge CP and Wiebe PD, *Crisis in Uganda: the breakdown of the health services*, pp. 25-41, Oxford: Pergamon.

Scoones I 1995. Investigating difference: applications of wealth ranking and household survey approaches among farming households in Southern Zimbabwe, *Development and Change*, 26:67-88.

Scott JC 1976. *The moral economy of the peasant*, New Haven: Yale University Press.

Seeley J and Kajura EB 1995. Grief and the community, In Sherr L, *Grief and AIDS*, pp. 73-85, Chichester: Wiley.

Seeley J, Wagner U, Mulemwa J and Kengeya-Kayondo J 1991. The development of a community-based HIV/AIDS counselling service in a rural area in Uganda, *AIDS Care*, 3:207-217.

Seeley J, Kajura E, Bachengana C, Okongo M, Wagner U, *et al* 1993. The extended family and support for people with AIDS in a rural population in Southwest Uganda: a safety net with holes?, *AIDS Care*, 5:117-122.

Seeley JA, Malamba S, Nunn A, Mulder D, Kengaya-Kayondo JF, *et al* 1994. Socio-economic status, gender and risk of HIV-1 infection in a rural community in South West Uganda, *Medical Anthropology Quarterly*, 8:78-89.

Seeley J, Nabaitu J, Taylor L, Kajura E, Bukenya T, *et al* 1996. Revealing gender differences through well-being ranking in Uganda, *PLA Notes*, February 1996:14-18.

Seligman CG 1966. *Races of Africa*, London: Oxford University Press.

Sen A 1981. *Poverty and famines: an essay on entitlement and deprivation*, Oxford: Clarendon Press.

Sen A 1997. *Development thinking at the beginning of the 21st century*, London: Suntory-Toyota International Centre for Economics and Related Disciplines.

Sen K 1994. *Ageing: debates on demographic transition and social policy*, London: Zed Books.

Sengendo J and Nambi J 1997. The psychological effect of orphanhood: a study of orphans in Rakai district, *Health Transition Review*, 7 (suppl):105-124.

Senteza-Kajubi W 1987. The historical background to the Uganda crisis, 1966-1986, In Wiebe PD and Dodge CP, *Beyond crisis: development issues in Uganda*, pp. 25-40, Kampala: Makerere Institute of Social Research and the African Studies Association.

Serwadda D, Sewankambo NK, Carswell JW, Bayley AC, Tedder RS, *et al* 1985. Slim disease: a new disease in Uganda and its association with HTLV-III infection, *The Lancet*, 2:849-852.

Serwadda D, Mhalu F, Karita E and Moses S 1994. HIV and AIDS in East Africa, In Essex M, Mbuop S, Kanki P and Kalengayi MR, *AIDS in Africa*, pp. 669-689, New York: Raven Press.

Serwadda D, Gray RH, Wawer MJ, Stallings RY, Sewankambo NK, *et al* 1995. The social dynamics of HIV transmission as reflected through discordant couples in rural Uganda, *AIDS*, 9:745-750.

Sewankambo NK, Carswell JW, Mugerwa RD, Lloyd G, Kataaha P, *et al* 1987. HIV infection through normal heterosexual contact in Uganda, *AIDS*, 1:113-116.

Shanin T 1987. *Peasants and peasant societies*, Oxford: Blackwell.

Shelton AJ 1965. Ibo aging and eldership: notes for gerontologists and others, *The Gerontologist*, 5:20-23.

Shelton AJ 1972. The Aged and eldership among the Igbo, In Cowgill DO and Holmes LD, *Aging and Modernisation*, pp. 31-49, New York: Meredith Corporation.

Silverman P and Maxwell RJ 1987. The significance of information and power in the comparative study of the aged, In Sokolovsky J, *Growing old in different societies: cross-cultural perspectives*, pp. 43-55, Acton: Copley.

Simmons LW 1945. *The role of the aged in primitive society*, New Haven: Yale University Press.

Singh N and Lawrence J 1997. *Promoting sustainable livelihoods: a briefing paper submitted to UNDP's executive committee*, Geneva: UNDP.

Southall A 1980. Social disorganisation in Uganda: before, during and after Amin, *Journal of Modern African Studies*, 18:627-656.

Southall A and Gutkind P 1957. *Townsmen in the Making: Kampala and its suburbs*, Kampala: East African Institute of Social Research.

Southwold M 1956. The inheritance of land in Buganda, *Uganda Journal*, 20:88-96.

Southwold M 1965. The Ganda of Uganda, In Gibbs JL, *Peoples of Africa*, pp. 83-118, New York: Holt, Rinehart and Winston Inc.

Spencer P 1965. *The Samburu*, London: Routledge and Kegan Paul.

Spradly JP 1979. *The Ethnographic Interview*, New York: Holt, Reinhart and Winston.

Spurr GB 1990. The impact of chronic undernutrition on physical work capacity and daily energy expenditure, In Harrison GA and Waterlow JC, *Diet and disease in developing countries*, pp. 24-61, Cambridge: Cambridge University Press.

Statistics Department 1992. *Integrated Household Survey of Uganda -1992. Manual of instructions to field workers*, Entebbe: Ministry of Finance and Economic Planning.

Statistics Department 1994. *Report on the Uganda National Integrated Households Survey 1992-3*, Entebbe: Ministry of Finance and Economic Planning.

Statistics Department 1995. *The 1991 Population and Housing Census*, Entebbe: Ministry of Finance and Economic Planning.

Statistics Department 1996a. *Demographic and Health Survey 1995*, Entebbe: Ministry of Finance and Economic Planning.

Statistics Department 1996b. *1996 Statistical Abstract*, Entebbe: Statistics department, Ministry of Finance and Economic Planning.

Sutnick AI, Lynch HT and Miller DG 1984. Cancer in Southeast Asia: strategies for control, *Journal of the American Medical Association*, 251:117-122.

Swift J 1989. Why are rural people vulnerable to famine?, *IDS Bulletin*, 20:8-15.

Tadria HMK 1987. Changes and continuities in the position of women in Uganda, In Wiebe PD and Dodge CP, *Beyond crisis: development issues in Uganda*, pp. 79-90, Kampala: Makerere Institute of Social Research.

Taylor L, Seeley J and Kajura E 1996. Informal care for illness in rural southwest Uganda: the central role that women play, *Health Transition Review*, 6:49-56.

Tembo G, Friesan H, Asiimwe-Okiror G, Moser R, Naamara W, *et al* 1994. Bed occupancy due to HIV/AIDS in an urban hospital medical ward in Uganda, *AIDS*, 8:1169-1171.

The Kabaka 1967. *The Desecration of my kingdom*, London: Constable.

The Kabaka 1971 [first publ. 1935]. Education, civilisation and "foreignisation" in Buganda (1935), In Low DA, *The mind of Buganda*, pp. 104-108, London: Heinemann.

The Regents of Buganda 1971 [first publ. 1900]. The 1900 negotiations, In Low DA, *The mind of Buganda: documents of the modern history of an African kingdom*, pp. 32-36, London: Heinemann.

Thomas J and Atkinson H 1995. *Participatory appraisal in Kibungo prefecture, Rwanda*, London: HelpAge International.

Thresh JM, Fishpool LDC, Otim-Nape GW and Fargette D 1994. African cassava mosaic virus disease: an under-estimated and unsolved problem, *Tropical Science*, 34:3-14.

Tlou S 1996. Empowering older women in AIDS prevention: the case of Botswana, *Southern African Journal of Gerontology*, 5:27-32.

Todaro MP 1989. *Economic development in the third world*, London: Longman.

Topouzis D and Hemrich G 1993. *The socio-economic impact of HIV and AIDS on rural families in Uganda: an emphasis on youth*, Geneva: UNDP.

Toulmin C 1992. *Cattle, Women and Wells: managing household survival in the Sahel*, Oxford: Clarendon Press.

Tout K 1989. *Ageing in Developing Countries*, Oxford: Oxford University Press.

Tout K 1993. A woman's place?, In Tout K, *Elderly Care: a world perspective*, pp. 289-296, London: Chapman Hall.

Townsend P 1962. The meaning of poverty, *The British Journal of Sociology*, 13:210-227.

Townsend P 1970. *The Concept of Poverty*, London: Heinemann.

Townsend P 1981. The structured dependency of the elderly: a creation of social policy in the twentieth century, *Ageing and Society*, 1:5-28.

Treas J and Logue B 1986. Economic development and the older population, *Population and Development Review*, 12:645-673.

Turnbull C 1972. *The Mountain People*, Simon and Schuster: New York.

Udvardy M and Cattell M 1992. Gender, Aging and Power in Sub-Saharan Africa: Challenges and Puzzles, *Journal of Cross-Cultural Gerontology*, 7:275-288.

Uganda Protectorate 1943. *Organisation of the Southwest labour migration routes*, Entebbe: Uganda Government.

UNAIDS 1998. *Uganda: epidemiological fact sheet on HIV/AIDS and sexually transmitted diseases*, Geneva: UNAIDS/World Health Organisation.

UNICEF 1990. *Children and AIDS: an impending calamity*, New York: UNICEF.

United Nations 1998. *Older women and support systems: new challenges E/CN.6/1998/4*, New York: United Nations.

United Nations 2002a. *Research Agenda on Ageing for the 21st century.* New York: United Nations.

United Nations 2002b. *International Plan of Action on Ageing 2002.* New York: United Nations.

United Nations 2002c. *Political declaration, World Assembly on Ageing, Madrid 12 April 2002.* New York: United Nations.

United Nations Development Program 1997. *Human development report,* New York: UNDP.

United Nations Population Division 1994. *World Population Prospects (the 1994 revision),* New York: United Nations.

Urassa M, Boerma JT, Ng'weshemi JZL, Isingo R, Schapink D, *et al* 1997. Orphanhood, child fostering and the AIDS epidemic in rural Tanzania, *Health Transition Review,* 7(Suppl 2):141-153.

Waite G 1988. The politics of disease: the AIDS virus and Africa, In Miller M and Rockwell RC, *AIDS in Africa: the social and policy impact,* pp. 145-164, Lewiston: Edwin Mellen Press.

Wallace CC and Weeks SG 1975. *Success or failure in rural Uganda: a study of young people,* Kampala: Makerere Institute of Social Research.

Wallis CAG 1953. *Report of an inquiry into African local government in the Protectorate of Uganda,* Entebbe: Government Printer.

Warnes AM 1994. Socio-economic change and the health of elderly people: future prospects for the developing world, In Phillips D and Verhasselt Y, *Health and Development,* pp. 156-167, London: Routledge.

West HW 1972. *Land policy in Buganda,* Cambridge: Cambridge University Press.

White GF, Bradley DJ and White AU 1972. *Drawers of water: domestic water use in East Africa,* Chicago: Chicago University Press.

Whitehead A 1990. Rural Women and Food Production in Sub-Saharan Africa, In Dreze J and Sen A, *The Political Economy of Hunger,* pp. 425-473, Oxford: Clarendon Press.

Whyte SR 1991. Medicines and self-help: the privatisation of health care in Eastern Uganda, In Hansen HB and Twaddle M (Eds), *Changing Uganda -The Dilemmas of Structural Adjustment and Revolutionary Change.* pp. 130-148, London: James Currey.

Wilk RR and Netting RM 1984. Households: changing forms and functions, In Netting RM, Wilk RR and Arnould EJ, *Households,* pp. 1-28, Berkeley: University of California Press.

Williams A, Bennett E, Himmavanh V and Salazar F 1996. "They just go home and die": health care and terminal illness in rural Northeast Thailand. *Asian Studies Review* 20(1):98-108.

Williams G and Tamale N 1991. *The Caring Community: coping with AIDS in urban Uganda,* London: ActionAid.

Wolf ER 1966. *Peasants,* Englewood Cliffs: Prentice-Hall.

World Bank 1990. *World Development Report 1990: Poverty,* Oxford: Oxford University Press.

World Bank 1993a. *Uganda: Growing Out of Poverty,* Washington DC: World Bank.

World Bank 1993b. *Uganda: social sectors,* Washington DC: World Bank.

World Health Organisation 1989. *Statement on AIDS and tuberculosis*, Geneva: World Health Organisation.

World Health Organisation 1991. *Report of the informal consultation on the needs of people with HIV infection and disease and their families*, Geneva: World Health Organisation.

World Health Organisation 2001. *Health and Ageing: a discussion paper*, Geneva: World Health Organisation.

World Health Organisation and UNICEF 1994. Uganda, In World Health Organisation, *Action for children affected by AIDS*, pp. 13-39, Geneva: World Health Organisation.

Wrigley CC 1957. Buganda: an outline economic history, *Economic History Review*, Second Series Vol 10:69-80.

Wrigley CC 1964. The changing economic structure of Buganda, In Fallers LA, *The King's Men*, pp. 16-63, London: Oxford University Press.

Wrigley CC 1970. *Crops and wealth in Uganda: a short agrarian history*, Oxford: Oxford University Press.

Yanagisako SJ 1979. Family and household: the analysis of domestic groups, *Annual Review of Anthropology*, 8:161-205.

Index